INFLUENCE OF STRUCTURAL

MOVEMENT ON SEDIMENTATION

DURING THE PENNSYLVANIAN

PERIOD IN WESTERN MISSOURI

INFLUENCE OF STRUCTURAL

MOVEMENT ON

SEDIMENTATION

DURING THE PENNSYLVANIAN

PERIOD IN WESTERN MISSOURI

RICHARD J. GENTILE

UNIVERSITY OF MISSOURI STUDIES VOLUME XLV

UNIVERSITY OF MISSOURI PRESS   COLUMBIA · MISSOURI

≋ Acknowledgments

I AM INDEBTED to numerous individuals and organizations whose help, understanding, and encouragement gave me the needed inspiration to complete this research project.

Dr. A. G. Unklesbay, at the time Chairman of the Department of Geology at the University of Missouri, Columbia, critically read the manuscript and willingly contributed many enlightening suggestions. His effort has made it possible to publish the completed manuscript.

Dr. Alfred C. Spreng, Professor of Geology at the University of Missouri at Rolla, offered much constructive criticism and consulted with me in the field.

I owe special thanks to Dr. Spreng and to Dr. Unklesbay for teaching me the basic principles and the modern concepts of stratigraphy when I was a student in their classes. The reader will notice that many of the stratigraphic concepts taught in their classrooms are incorporated in this report. I acknowledge their help and guidance with this expression of my deep appreciation.

Dr. Walter V. Searight, Principal Geologist of the

Acknowledgments

Missouri Geological Survey and Water Resources, now retired, considered the project area as one of particular stratigraphic importance and suggested that detailed geologic mapping should be done in the area.

Dr. Wallace B. Howe, Assistant Director of the Missouri Geological Survey and Water Resources, and Mrs. Mary H. McCracken, Research Geologist for the Survey, gave freely and generously of the knowledge that they possess concerning the project area, and I wish to recognize here their generosity.

I am deeply indebted to Professor Thomas R. Beveridge, former Director of the Missouri Geological Survey and Water Resources, and to Dr. William C. Hayes, the present Director, for providing the financial assistance and for making available the personnel and facilities of the Survey. The field work was done from 1960 to 1965, while I was in their employ. I am especially grateful to Douglas R. Stark, Chief Draftsman, for drafting the illustrations and to Glenda Otis for typing the manuscript.

Numerous people of western Missouri allowed access to mining operations or rock outcrops on their property. Their kindness facilitated my investigations, and I am pleased to acknowledge my debt to them.

I also wish to thank the United States Air Force for permitting me to study the excavations during construction of Minuteman missile sites and for releasing to the Missouri Geological Survey and Water Resources the test cores drilled at these sites. The information obtained from a study of these cores has proved invaluable.

R. J. G.

University of Missouri — Kansas City
January, 1967

Contents

List of Figures

List of Plates
(In separate pocket)

≋≋ Introduction

PENNYSLVANIAN STRATA in the midcontinent and eastern interior regions of the United States are composed of relatively thin beds of limestone, sandstone, shale, coal, and underclay. These occur in cyclical sequences or "cyclothems." The majority of the units, especially the limestone, coal, underclay, and some shale units, are considered to be persistent by most geologists and to occur over large areas. Persistent lithologic units a few inches thick have been mapped over areas of many hundreds of square miles. Nevertheless, in western Missouri a few prominent lithologic units of some of the cyclothems pinch out or disappear, while other units are characterized by abrupt variations in lithology and thickness.

It is believed that irregularities in the depositional surface contemporaneous with sedimentation affected the persistence, lithology, and thickness of these units. Field and subsurface mapping have disclosed several northwestward-trending structures. When mapped in the field these structures appear as broad, low, asymmetrical folds with steeply dipping southwestern limbs. The steeply dipping southwestern limb is in many areas of western Missouri truncated by high-angle faulting. When projected into the subsurface the faulting can be recognized in Precambrian basement rocks. The faults appear to bound a northwest-ward-aligned series of tectonic blocks.

One of the largest of the northwestward-trending structures is the Schell City-Rich Hill anticline (Pl. 1 and Fig. 13). Faulting along the southwestern limb of this structure has been recognized in field studies of rocks belonging to the Cherokee and Marmaton groups of the Desmoinesian Series and in subsurface studies of lower Paleozoic and Precambrian rocks.

Although other processes were operative, it is postulated that differential movement of deep-seated tectonic blocks contemporaneous with sedimentation influenced the depositional cycle. The variation in physical characteristics of many thin sedimentary units, when traced to the vicinity of the faulted southwestern limb of the Schell City-Rich Hill anticline, suggests that movements associated with this structure strongly influenced sedimentary processes. The variations in physical characteristics of the lithologic units persist for many miles into areas of comparatively little structural disturbance, although low rank metamorphism is recognized in the shale units in the faulted area. The complete sequence of strata comprising the Cherokee and Marmaton groups was studied in detail, but only rocks belonging to the upper Cherokee and lower Marmaton groups are included in this report. The remainder of strata comprising these two groups are not included, because of the discontinuous nature of the units and the difficulty in tracing key beds across the area. Exact correlation of units is prerequisite to a study of this nature.

Rocks belonging to the upper Cherokee and lower

Marmaton groups are exposed in a belt that crosses Missouri in a northeast-southwest direction. The outcrop belt of these rocks in western Missouri ranges in width from 15 to 20 miles. The average total thickness of strata is about 200 feet. That portion of the belt from the Kansas-Missouri line to a few miles east of the Bates-Henry county line, an outcrop distance of 45 miles, will be discussed in this report (Fig. 1). It includes most of Bates County, northwestern Vernon County, and a small portion of western Henry County. This region will be called *the Bates County area* in this study.

The investigation included a study of several hundred rock outcrops and artificial exposures. Considerable subsurface data were used, particularly in areas where cores were made available by the construction of a complex system of Minuteman missile sites by the United States Air Force.

FIGURE 1. Index map of Missouri, showing outcrop belt of the Cherokee and Marmaton groups. Bates County area shown by solid shading.

≋≋ Descriptive Stratigraphy

Pennsylvanian System

The exposed rocks of the Bates County area of western Missouri belong to the Desmoinesian Series of the Pennsylvanian System. The Cherokee and Marmaton groups are the major rock groups that represent this series in Missouri. The subdivision of the Desmoinesian Series, as recognized by the Missouri Geological Survey (Searight, 1961, pp. 81-95), is shown in Figure 2. The rocks from the top of the Mineral Formation upward to the top of the Altamont Formation are discussed in this report.

ROBINSON BRANCH FORMATION

The Robinson Branch Formation includes the beds from the top of the Mineral coal to the top of the Robinson Branch coal. The formation averages about 10 feet in thickness, but locally in east-central Bates County it thins to a few inches (Pl. 2).

The lower part of the Robinson Branch Formation is quite variable, lithologically, from place to place. Four different lithologic types have been recognized directly overlying the Mineral coal and include: (1)

Pennsylvanian System				
Missourian Series		Pleasanton Group		Unnamed Formation

Desmoinesian Series	Cygnian Stage	Marmaton Group	Appanoose Subgroup	Holdenville Formation
				Lenapah Formation
				Nowata Formation
				Altamont Formation
				  Worland Member
				  Lake Neosho Member
				  Amoret Member
				Bandera Formation
				  Bandera Quarry Member
				  (Mulberry coal)
				Pawnee Formation
				  Coal City Member
				  Mine Creek Member
				  Myrick Station Member
				Labette Formation
				  (Lexington coal)
				  (Alvis coal)

			Fort Scott Subgroup	Higginsville Formation
				Little Osage Formation
				  Houx Member
				  (Summit coal)
				Blackjack Creek Formation

		Cherokee Group	Cabaniss Subgroup	Excello Formation
				Mulky Formation
				  Breezy Hill Member
				Lagonda Formation
				  (Squirrel sandstone)
				Bevier Formation
				Verdigris Formation
				  (Wheeler coal)
				  Ardmore (Verdigris) Member
				Croweburg Formation
				Fleming Formation
				Robinson Branch Formation
				Mineral Formation
				Scammon Formation
				Tebo Formation
				Weir Formation

	Venteran Stage		Krebs Subgroup	Seville Formation Bluejacket Formation Drywood Formation Rowe Formation Warner Formation Hartshorne (?) Formation
Atokan Series				Riverton Formation

FIGURE 2.   Subdivisions of the Desmoinesian Series in the mid-continent area.

black, fissile, noncalcareous shale containing phosphatic concretions (Pl. 2, Sec. 10); (2) black, platy, calcareous shale (Pl. 2, Sec. 1); (3) dark-gray to black, thin-bedded, argillaceous limestone containing abundant *Desmoinesia muricatina* (Dunbar and Condra) (Pl. 2, Secs. 7,[1] 12); and (4) dark-gray, noncalcareous, flaky shale (Pl. 2, Sec. 8). Lithologic types designated (2) and (3) have been observed to grade into each other laterally and have the characteristic of grading vertically into the top of the Mineral coal. The Mineral coal is the top unit of the underlying Mineral Formation. It is over 4 feet thick in the Rich Hill district and was extensively mined before the turn of the century. In former times this coal was known as the lower Rich Hill.

A few inches to several feet of dark-gray flaky shale overlies one of the four lithologic types previously mentioned. Near the center of this dark-gray flaky shale is a thin zone of limestone nodules mixed with clay matrix (Pl. 2, Sec. 2). This stratigraphic position is occupied by a sandstone bed in western Henry County (Pl. 2, Sec. 10) and in the subsurface in the vicinity of Rich Hill (Pl. 2, Sec. 4). It appears that the nodules of limestone grade laterally into the sandstone.

The Robinson Branch coal is the upper unit of the formation. It is underlain by an underclay that is poorly developed locally. The Robinson Branch coal

[1]Detailed written descriptions and localities of the numbered stratigraphic sections referred to in the text portion and illustrated in the cross-profiles are available in the appendix of the unpublished doctoral dissertation entitled "Stratigraphy Sedimentation and Structure of the Upper Cherokee and Lower Marmaton (Pennsylvanian) Rocks of Bates County, and Portions of Henry and Vernon Counties, Missouri" by Richard J. Gentile, 1965. Copies are on file at the University of Missouri at Rolla and the Missouri Geological Survey and Water Resources, Rolla, Missouri.

varies considerably in thickness throughout western Missouri. It is 2 feet thick in the Rich Hill district (Pl. 2, Sec. 3) and is strip-mined in tandem with the Mineral coal. The Robinson Branch coal was formerly known as the upper Rich Hill. Variations in thickness from 2 feet to a smut have been observed in a lateral distance of a few feet along the highwall of coal strip mines (Pl. 2, Sec. 1). The Robinson Branch coal commonly is coated with the white fibrous mineral melanterite (Fe SO_4. $7H_2O$).

FIGURE 3.   Sandstone lens near the middle of the Robinson Branch Formation, western Henry County SE¼ SW¼ sec. 19, T. 41 N., R. 28 W. Pick has been placed at the base of the Robinson Branch coal bed.

The Robinson Branch Formation is overlain by the Fleming Formation and underlain by the Mineral Formation. All three formations are cyclic sequences that are in many ways lithologically and paleontologically similar to each other. The lower boundary of the Robinson Branch Formation, which has been placed at the top of the Mineral coal, is relatively sharp. The Mineral coal has been recognized to persist throughout most of the outcrop distance from Kansas to western Henry County, although the thickness varies considerably from place to place.

The upper boundary of the Robinson Branch Formation lies at the top of the Robinson Branch coal and below the Fleming Formation. Northeastward along the outcrop belt from Rich Hill the Robinson Branch coal thins to a smut or in some places is absent. At these places the stratigraphic boundary must be inferred from the position of the associated strata.

FLEMING FORMATION

The Fleming Formation includes beds from the top of the Robinson Branch coal bed to the top of the Fleming coal bed.

The formation is about 7 feet thick in northwestern Vernon County, but it thickens to more than 45 feet in the Rich Hill district of south-central Bates County (Pl. 2, Sec. 4). Northeastward from Rich Hill it thins irregularly and is about 5 feet thick in western Henry County.

The bottom part of the Fleming Formation is composed of several inches to 2 or 3 feet of black platy shale, which grades into thin-bedded argillaceous limestone containing abundant *Desmoinesia muricatina* (Dunbar and Condra). Other fossils include *Lingula,* bryozoans, echinoid spines, and sparse gastropods. The black platy shale sometimes grades downward into coal. This characteristic makes it difficult to remove it from the coal during mining operations. The strata of this position are quite similar in places to those overlying the Mineral coal and may be confused, with the result that the Robinson Branch coal is miscorrelated with the Mineral. In northwestern Vernon County a few inches of black shale containing irregular-shaped phosphatic concretions overlies the Robinson Branch coal (Pl. 2, Secs. 1, 2).

The middle part of the Fleming Formation consists of soft, dark-gray shale that in north-central Vernon County contains abundant flattened siderite concretions some 6 inches in diameter. A thick, lenticular sandstone bed occupies this interval at Rich Hill (Pl. 2, Sec. 4).

The Fleming coal, which is the top of the Fleming Formation, is poorly developed and is represented in most places in the Bates County area of western Missouri by a coal smut or a thin bed of dark-gray shale. The greatest observed thickness was 3 inches in northwestern Vernon County. The Fleming coal, where present, is underlain by a few inches of underclay.

The Fleming Formation is underlain by the Robinson Branch Formation and is overlain by the Croweburg Formation. The lower boundary is placed on top of the Robinson Branch coal bed or coal horizon, which is present in most areas. The upper boundary is the top of the Fleming coal bed. The Fleming coal bed is not present in many outcrops, and where it is absent the upper boundary must be inferred from the position of the associated strata.

CROWEBURG FORMATION

The Croweburg Formation includes beds from the top of the Fleming coal to the top of the Croweburg coal. The formation varies in thickness from 7 to 20 feet and averages about 10 feet (Pl. 2). It is thickest in south-central Bates County, where the formation includes several lenticular sandstone and shale beds (Pl. 2, Secs. 4, 5).

In most places the lower part of the Croweburg Formation consists of dark shales that grade laterally and vertically into thin limestone beds that are similar lithologically and faunally to the limestone beds overlying the Mineral and Robinson Branch coal beds. These limestone beds are dark-gray, thin-bedded, argillaceous, and contain abundant *Desmoinesia muricatina* (Dunbar and Condra). The limestones are ferruginous, and they weather to a dark reddish brown. Other fossils observed in the lower part of the Croweburg Formation include *Crurithyris,* sparse numbers of linoproductids, and fragments of other productids.

Lenses of light-gray sandstone 2 to 10 feet thick overlie the dark shales and thin limestone beds in the Rich Hill district of south-central Bates County (Pl. 2, Secs. 3, 4, 5). The contact between the sandstone and underlying shale is sharp in most places. The sandstone is micaceous, cross-bedded, and locally calcareous. The top surface is pitted with holes about one-half inch in diameter and 1 inch deep. These features appear to have contained fossil root material that has been weathered away. When traced along the outcrop belt from the Rich Hill district to western Henry County and northwestern Vernon County, the sandstone grades into greenish-gray clay with nodules of limestone. The sandstone grades upward into shale or into the underclay of the Croweburg coal bed. The Croweburg coal bed is very persistent and is about 1 foot thick. Because of its uniform thickness it has been informally named by miners the "one-foot" or "ten-inch" coal bed. Nevertheless, in a few isolated outcrops the Croweburg coal is characterized by variations in thickness and may become thin-bedded and contain interbedded shale lenses.

The Croweburg Formation is underlain by the Fleming Formation and overlain by the Verdigris Formation. The upper boundary is placed at the top of the Croweburg coal bed and the lower boundary on top of the Fleming coal bed. In places where the Fleming coal is absent, the boundary is placed slightly below the lowest thin limestone bed in what is believed to be the approximate stratigraphic position of the Fleming coal bed.

Verdigris Formation

The Verdigris Formation includes the beds from the top of the Croweburg coal bed to the top of the Wheeler coal bed. The Verdigris limestone comprises the beds called Ardmore by Gordon (1893, p. 20-21) in Macon County, Missouri. Although the name Ardmore has precedence, the name Verdigris has had wider usage and for that reason was adopted at the Nevada Conference (Searight, 1953, p. 2748). In the Bates County area of western Missouri the Verdigris limestone has been called the Rich Hill limestone and the "sump" rock.

The Wheeler coal bed is the top unit of the Verdigris Formation. The Wheeler coal was called the Williams coal in Vernon and Bates counties (Greene and Pond, 1926, p. 52).

The Verdigris Formation is 45 feet thick in south-central and southeastern Bates County. It thins along the outcrop belt to about 20 feet in northwestern Vernon County and is less than 15 feet thick in western Henry County (Pl. 3).

Lithologically, the Verdigris Formation is extremely

variable. The lower part is composed of several feet of gray flaky shale that, when traced northeastward along the outcrop belt from northwestern Vernon County, thickens abruptly in the vicinity of Rich Hill and the top part of the shale grades laterally into 6 feet or more of fine-grained, micaceous sandstone (Pl. 3, Secs. 14, 15). The sandstone grades upward into 2 or 3 inches of sandy to nodular limestone containing abundant fragments of brachiopods of the genera *Mesolobus, Desmoinesia, Chonetinella,* and *Eolis-sochonetes* (Pl. 3, Secs. 5[2], 14). Traced southwestward and northeastward along the outcrop belt from south-central Bates County and north-central Vernon County, the sandy or nodular limestone unit grades laterally into thin-bedded to blocky, dark-gray limestone containing *Mesolobus mesolobus* and *Des-moinesia*. It is overlain by about 2 or 3 feet of black fissile shale that contains abundant phosphatic concretions (Pl. 3, Sec. 14). The black fissile shale persists from the vicinity of Rich Hill southwestward to beyond the Kansas State boundary (Pl. 3, Sec. 1). Northeastward from Rich Hill to at least as far as western Henry County this interval is occupied by a few inches of greenish-gray shale with irregular-shaped, phosphatic concretions.

The Verdigris limestone consists of two beds separated by shale from approximately two miles southwest of Rich Hill to the Kansas State boundary (Pl. 3, Secs. 1, 13). The lower bed is dark-gray, jointed, and usually occurs as one bed that characteristically weathers to large rectangular blocks. Locally it has

[2]Most stratigraphic sections shown in the cross-profiles contain only a part of the section as measured in the field. This method of illustration was necessary in order to present lithologic detail. Consequently, many numbers referring to stratigraphic sections appear on more than one profile.

been called the "diamond" rock. The upper bed is light-gray, wavy-bedded, and contains *Desmoinesia* and other brachiopods. The Verdigris limestone is represented in south-central Bates County and north-central Vernon County by over 5 feet of light-gray, thick-bedded, nodular limestone that in some places contains numerous thin thanatocoenoses of fossil fragments belonging mostly to the genera *Mesolobus* and *Eolissochonetes* (Pl. 3, Sec. 15).

FIGURE 4.   Phosphatic shale bed underlying Verdigris limestone, north-central Vernon County, NE¼ sec. 34, T. 38 N., R. 31 W. Head of pick has been placed near the base of the phosphatic shale bed. Phosphate occurs as thin lenses and concretions in a greenish-gray shale. Southwestward from Rich Hill this interval grades abruptly laterally into 2 feet or more of black fissile shale. Light-colored beds below head of pick are composed of sandy shale.

The Verdigris limestone thins in northeastern Bates and west-central Henry County and is from 6 inches to 3 feet thick. Locally it is bedded and contains some sand, but in most places it is represented by a rubble of nodules of limestone (Pl. 3, Secs. 11, 18).

FIGURE 5.   Nodular bedding in the Verdigris limestone, north-central Vernon County, NE¼ sec. 34, T. 38 N., R. 31 W.

The Wheeler coal is separated from the Verdigris limestone by clay and underclay. The coal is 6 inches thick or more, approximately 2 miles southwest of Rich Hill (Pl. 3, Sec. 13), and this thickness is believed to persist to the Kansas-Missouri boundary. The Wheeler coal and underclay are not present in many places northeast along the outcrop belt from Rich Hill, but in western Henry County it thickens to 6 inches or more (Pl. 3, Sec. 11).

Sandstone cuttings from well borings have been obtained from the interval between the Verdigris limestone and the Wheeler coal in extreme southwestern Bates County. It is assumed that a sandstone bed is present at this position in the area.

The lower boundary of the Verdigris Formation is easily recognized because in surface exposures, as well as in drill cuttings, the Croweburg coal bed is (1) persistent in extent, (2) fairly uniform in thickness, and, most important, (3) it is the first coal bed below the Verdigris limestone.

The upper boundary of the Verdigris Formation is difficult to recognize in some areas in south-central Bates County because the Wheeler coal bed, with its associated underclay and thin-bedded cap limestone, is absent, and this interval contains gray shales. At these places the boundary between the Verdigris and the Bevier formations is placed within the shale sequence at the approximate stratigraphic position of the top of the Wheeler coal. However, at some exposures the thin-bedded cap limestone that overlies the Wheeler coal lies almost directly on the Verdigris limestone, and the Wheeler coal and underclay have all but pinched out; the boundary between the Verdigris and Bevier formations is placed for all practical purposes on top of the Verdigris limestone (Pl. 3, Secs. 5, 16).

FIGURE 6.   Wheeler coal bed and associated strata, southwestern Bates County, SE¼ sec. 6, T. 39 N., R. 32 W. Coal is overlain by approximately 2 feet of dark-gray shale (position of pick handle). A thin abundantly fossiliferous limestone bed overlies the shale. Talus composed of Lagonda shale covers the slope above the thin limestone bed. The Wheeler coal bed is 9 inches thick and is separated from the Verdigris limestone (not shown) by approximately 2 feet of underclay. A few miles east of this location the strata between the Verdigris limestone and the thin limestone bed shown above are considerably reduced in thickness, or they are absent.

BEVIER-LAGONDA FORMATION

In this study the Bevier and Lagonda formations will be discussed as one unit because the Bevier coal bed has not been recognized in the Bates County area of western Missouri; as a result the upper boundary of the Bevier Formation could not be established. The Bevier-Lagonda Formation includes beds from the top of the Wheeler coal bed to the top of the lowermost of three thin coal beds or smuts that lie below the Mulky underclay in Henry County. Hover (1958, p. 44), working in west-central Henry County, tentatively correlated the Iron Post coal of Oklahoma and Kansas with a coal bed that is possibly the lowermost of this sequence in Henry County. These coals appear at approximately the same stratigraphic position in the two areas.

The exact thickness of the Bevier-Lagonda Formation could not be determined in the western part of the area because the upper part of the formation is quite variable lithologically, and consequently the boundary could not be established (see Pl. 4). Nevertheless, it is doubtful if the formation is over 50 feet or less than 25 feet thick.

The Bevier-Lagonda Formation is composed predominantly of shale and sandstone, with minor amounts of coal and thin-bedded, argillaceous limestone. A few inches of dark-gray, thin-bedded limestone that weathers to reddish-brown, earthy blocks is present at the base of the formation and locally overlies the Wheeler coal bed or is separated from it by a few inches to a foot or two of dark shale (Pl. 4, Secs. 5, 21). This limestone contains *Desmoinesia* and sparse *Linoproductus*.

A 3-inch bed of black fissile shale that is surrounded

by gray shale lies 10 to 15 feet above the base of the formation in western Henry County and northeastern Bates County (Pl. 4, Secs. 18, 26). It is assumed that this unit is associated with the Bevier coal, but its exact stratigraphic position is not known. A few inches to 2 feet of dark-gray, thin-bedded, arenaceous limestone with abundant *Mesolobus* occupies the same stratigraphic position throughout most of Bates County and northwestern Vernon County (Pl. 4, Secs. 5, 21, 22).

A thin bed of coal lies at the top of the Bevier-Lagonda Formation in west-central Henry County and northeastern Bates County (Pl. 4, Secs. 11, 25). As previously mentioned, this coal has been tentatively correlated with the Iron Post coal of Oklahoma (Hover, 1958, p. 44). Nevertheless, it could not be traced with any degree of certainty southwestward from northeastern Bates County. A coal horizon that is marked by an arenaceous underclay and a thin-bedded limestone is present in south-central Bates County (Pl. 4, Sec. 21). A correlation between these strata and the coal bed, tentatively called the Iron Post by Hover, in northeastern Bates County is, however, a matter of conjecture.

Several feet of sandstone and sandy shale underlie the Iron Post (?) coal in northeastern Bates and western Henry counties (Pl. 4, Secs. 11, 18, 25). The sandstone has been informally called by drillers the "Squirrel" sandstone because of its tendency to "jump around" from one stratigraphic position to another in the interval between the Ardmore limestone and the Blackjack Creek limestone. This is a valid deduction because the "Squirrel" sandstone grades laterally into shale in comparatively short distances and may occur at more than one stratigraphic horizon. At most outcrops in the Bates County area of western Missouri two or more sandstones occur in the interval between the Verdigris and Blackjack Creek limestones (Pl. 4, Secs. 11, 21, 22, 25).

The Bevier-Lagonda Formation is underlain by the Verdigris Formation and overlain by the Mulky Formation. Both boundaries are problematic in most areas.

Mulky Formation

Present usage of the Missouri Geological Survey (Searight, 1961, p. 89) places the lower boundary at the top of the lowermost coal (Iron Post [?]) in a succession of three coal beds or horizons lying below the Mulky underclay in Henry County, Missouri. The upper boundary of the formation is placed at the top of the Mulky coal bed.

The Mulky Formation is 25 to 35 feet thick in northeastern Bates County and western Henry County (Pl. 4). Southward along the outcrop from this area to the Kansas line, the thickness could not be determined with any degree of accuracy because the lower boundary could not be recognized.

The lower part of the Mulky Formation in northeastern Bates and western Henry counties is composed of several feet of limestone and interbedded shale. The limestone beds are thin-bedded, argillaceous, medium- to dark-gray, and contain small amounts of sand (Pl. 4, Secs. 11, 18). The thickness of the individual limestone beds varies from less than an inch to over 1 foot. The thicker beds have the characteristic of weathering to large rectangular slabs. Fossils include fusulinids, *Mesolobus*, *Antiquatonia*, *Kozlowskia*, abundant small crinoid columnals, and coiled gastropods. The shale lying between the limestone beds is medium-gray and slightly sandy. In some places poorly developed underclays are present below one or more of the limestone beds (Pt. 4, Sec. 25). The limestone and interbedded shales are not well developed southwest of northeastern Bates County. At several localities in south-central Bates County (Pl. 4, Secs. 21, 22) discontinuous thin-bedded limestone with associated underclay has been recognized, but it could not be determined which limestone, if any, was its correlative in northeastern Bates County. It is assumed that the well-developed coal horizons at this interval in northeastern Bates and western Henry counties are localized in areal extent throughout the remainder of the Bates County area. The lower part of the Mulky Formation is not well exposed in northwestern Vernon County.

The "Squirrel" sandstone of drillers' terminology is present in the upper part of the formation. It is a fine-grained, micaceous sandstone several feet thick. The top part in some places becomes calcareous and grades into nodular limestone mixed with greenish-gray clay. The term Breezy Hill was formerly restricted to this nodular limestone.

The Mulky coal is the top unit of the Mulky Formation. It is 2 feet or more in thickness in western Henry County (Pl. 4, Sec. 18) and has been mined in small operations. Southwestward from the Bates-Henry county boundary the Mulky coal thins to a smut in most areas of Bates County. The Mulky coal in northwestern Vernon County thickens locally to 1 foot or more and contains a well-developed underclay (Pl. 4, Sec. 20).

The Mulky Formation is underlain by the Lagonda Formation and overlain by the Excello Formation.

The upper boundary is easily determined, since the Mulky coal or associated strata are present in most places. The lower boundary is difficult to place, because of the variations in lithology at this position in most of the Bates County area.

EXCELLO FORMATION

The Excello Formation consists of a thin shale unit lying between the top of the Mulky coal and the base of the Blackjack Creek limestone. The Excello Formation is about 3 feet thick in northwestern Vernon County and thins to less than 1 foot in most areas of Bates County. It increases in thickness to 5 feet or more in western Henry County (Pl. 4).

The Excello Formation in northwestern Vernon County consists of black fissile shale overlain by dark-gray shale that grades upward into light-gray shale (Pl. 4, Secs. 19, 20). The black fissile shale contains small irregular-shaped to round phosphatic concretions. In south-central Bates County, the Excello Formation thins to less than 1 foot and is composed of black to dark-gray soft shale that weathers greenish-gray (Pl. 4, Sec. 22). The shale loses its fissility, and the phosphatic concretions are smaller and more irregular in shape. Locally in this area less than 1 inch of dark-gray fossiliferous limestone overlies the Mulky coal or coal smut.

The Excello Formation thickens in the vicinity of the Bates-Henry County border and is lithologically similar to its correlative in northwestern Vernon County. It includes 3 feet or more of black fissile shale overlain by dark-gray shale that grades upward into light-gray shale (Pl. 4, Sec. 18). The black fissile shale contains irregular-shaped to round phosphatic concretions and large, finely crystalline, fossiliferous limestone concretions, many of which are spheroidal and 2 feet or more in diameter. Conodonts and a few low-spired crushed gastropods have been observed in the black, fissile shales.

The Excello Formation is underlain by the Mulky Formation and overlain by the Blackjack Creek Formation. Both upper and lower boundaries are relatively sharp.

BLACKJACK CREEK FORMATION

The Blackjack Creek is the lowermost formation of the Marmaton Group. The Blackjack Creek Formation includes the beds between the top of the Excello Formation and the base of the Little Osage Formation. The formation varies irregularly in thickness from 3 to 7 feet (Pl. 5).

The Blackjack Creek is composed of limestone and shale, with limestone predominating. The lower part of the formation commonly consists of finely crystalline, bedded limestone that weathers brown or tan. The limestone in most exposures is jointed with the main joint pattern oriented SW-NE and NW-SE. As a result of this jointing the limestone commonly weathers out in diamond-shaped blocks. Locally the lower part may become unevenly bedded or form a thick, nodular bed (Pl. 5, Sec. 30). Fossils include *Composita, Neospirifer, Mesolobus* and other brachiopods, horn corals, fusulinids, and abundant crinoid columnals.

The upper part of the Blackjack Creek Formation in northwestern Vernon County consists of thick-bedded, blocky limestone that contains abundant *Chaetetes* colonies near the top (Pl. 5, Sec. 27). When traced northeastward along the outcrop belt from northwestern Vernon County into Bates and Henry counties, the upper part of the formation becomes nodular-bedded and in most places consists of calcareous shale mixed with argillaceous limestone nodules. These weather to a coarse rubble.

The upper and lower parts of the Blackjack Creek are separated by shale in northeastern Bates County (Pl. 5, Sec. 34). The lower boundary is a relatively sharp contact between limestone of the Blackjack Creek and shale of the Excello Formation. The upper boundary in some places is gradational between nodular limestone of the Blackjack Creek Formation and shale of the Little Osage Formation (Pl. 5, Secs. 5, 33, 34).

LITTLE OSAGE FORMATION

The Little Osage Formation includes the beds between the top of the Blackjack Creek and the base of the Higginsville Formation (Pl. 5). The Little Osage Formation includes among other beds the Houx Member and the Summit coal. The formation varies irregularly in thickness from 2½ to 11 feet along the outcrop belt from northwestern Vernon County to western Henry County (Pl. 5).

The lower part of the Little Osage Formation consists of gray shale overlain by a poorly developed thin underclay. The Summit coal is represented by a thin coal smut or is absent throughout most of the area. The greatest observed thickness of the Summit coal bed is in west-central Bates County in the vicinity of the Marais des Cygnes River, where it is 4 inches thick (Pl. 5, Sec. 31). The Summit coal is overlain by a few inches of dark-gray to black shale that is commonly interbedded with dark-gray, shaly, argillaceous limestone. The shales and limestones contain *Kozlowskia splendens* (Norwood and Pratten), *Derbyia crassa* (Meek and Hayden), *Composita,* and *Mesolobus mesolobus* (Norwood and Pratten).

The most persistent unit in the Little Osage Formation is a black fissile shale that averages about 1½ feet in thickness. It contains numerous flattened to round phosphatic concretions. The black fissile shale is overlain by a few inches to a few feet of shale that is black at the bottom and grades upward into light-gray shale at the top. At places where the shale is thickest it commonly contains several thin lenses of limestone that are darkest in color near the bottom. The limestones are shaly to blocky, argillaceous, and contain *Kozlowskia, Crurithyrus, Mesolobus mesolobus* (Norwood and Pratten), and *Antiquatonia* (Pl. 5, Sec. 31). These limestone lenses occupy the same stratigraphic position and are lithologically and faunally similar to the Houx limestone of Johnson County and counties to the northeast of Johnson. The Little Osage Formation is easily recognizable in outcrop and in the subsurface because, lithologically, it is predominantly shale and occupies the interval between two formations that are predominantly limestone.

HIGGINSVILLE FORMATION

The Higginsville Formation contains only one unit, the Higginsville limestone. The formation varies irregularly in thickness from 12 to 21 feet along the outcrop belt from northwestern Vernon to western Henry County (Pl. 5). It is approximately 20 feet thick in central Bates County, southwest of Butler (Pl. 5, Sec. 32), and 21 feet thick in southwestern Bates

County at the town of Pleasant Gap (Pl. 5, Sec. 33). Both areas of increased thickness appear to be local and of limited areal extent.

The Higginsville Formation is composed of limestone. The lower part of the formation is even to wavy-bedded and becomes crinkly-bedded to nodular at the top. The limestone is light-gray with dark-gray mottling. Minor amounts of greenish-gray clay occur between bedding planes and joints. Small waxy-looking nodules of brown chert are a minor constituent. Fossils include *Phricodothyris, Composita, Crurithyris,* abundant crinoid columnals, and sparse horn corals. Brachiopod fossils are in most places recrystallized and held firmly in the limestone matrix. The upper part of the formation contains abundant fusulinids and *Chaetetes* colonies. The *Chaetetes* colonies form mounds on the top surface of the formation.

FIGURE 7.   Wavy-bedded limestone in basal part of the Higginsville Formation, northeastern Bates County, NW¼ NW¼ sec. 12, T. 41 N., R. 30 W. Re-entrant is formed by erosion of the underlying soft shale of the Little Osage Formation. Right foot of man rests on top of black fissile shale bed.

The Higginsville Formation is underlain by the Little Osage Formation and overlain by the Labette Formation. The Labette Formation lies conformably on the Higginsville Formation in northwestern Vernon County (Pl. 5, Sec. 28). At this location sandstone-filled shallow channels are present in the top part of the formation.

Several thin limestone beds occur locally at the base of the Higginsville (Pl. 5, Sec. 31). The limestone beds at the bottom of this sequence resemble those of the Little Osage Formation in that they are dark-gray, thin-bedded, and fossiliferous, while those at the top are light- to medium-gray, sparsely fossiliferous, and resemble the limestone of the Higginsville Formation. The limestone beds in the middle are gradational, and the contact between the Little Osage and the Higginsville formations has to be placed according to the judgment of the geologist.

LABETTE FORMATION

The Labette Formation includes the beds from the top of the Higginsville Limestone Formation to the base of the Anna Shale Member. The Labette varies considerably in thickness from place to place along the outcrop belt from northwestern Vernon County to western Henry County. The minimum thickness is about 25 feet, and the maximum thickness is over 45 feet (Pl. 6). The average thickness is approximately 30 feet.

The Labette Formation is composed predominantly of sandstone, shale, arenaceous limestone, calcareous sandstone, and minor amounts of limestone and conglomerate. The formation includes two coal beds or horizons: a lower coal (Alvis) and an upper coal

(Lexington). The Alvis is from 1 to 6 inches thick and is separated in Bates County and western Henry County from the Higginsville limestone by about 2 feet of gray underclay. In northwestern Vernon County 5 feet or more of sandstone occurs below this coal (Pl. 6, Sec. 36). The sandstone is fine-grained, light-gray, massive, and cross-bedded to thin-bedded. Locally, it fills channels eroded into the top of the Higginsville limestone. A thin conglomerate composed of particles of limestone lies below the coal in northwestern Vernon County and appears to be of very limited extent (Pl. 6, Sec. 37).

The Alvis coal is overlain by 1 to 5 feet of dark-gray to black calcareous shales that in most exposures contain from one to several thin beds of argillaceous limestone that are gradational with the shale. The shales

FIGURE 8.   Conglomerate lens underlying Alvis coal, northwestern Vernon County, NW¼ NE¼ sec. 26, T. 38 N., R. 33 W. Head of pick has been placed at base of coal. Conglomerate appears to have filled in a small channel eroded into the underclay of the Alvis coal. Channel-filling sandstone occupies the interval in most of northwestern Vernon County.

and argillaceous limestones contain abundant brachiopod fossils. In some exposures brachiopod fossils form thin beds of coquina. Brachiopod fossils include *Mesolobus mesolobus* (Norwood and Pratten), *Antiquatonia, Composita, Derbyia crassa* (Meek and Hayden), *Neospirifer, Kozlowskia, Linoproductus,* and *Neochonetes.* Other fossils are crinoid columnals and fenestellate bryozoans. As a rule, individual genera and species of brachiopods are restricted to zones (teilzones or local range zones), and as a result some coquina beds are composed almost entirely of *Mesolobus mesolobus* while other coquina beds contain *Derbyia* or *Composita* as the dominant genus.

The shale overlying the Alvis coal grades upward into about 20 feet of sandstone and shale, which become very calcareous in southwestern Bates and northwestern Vernon counties (Pl. 6, Sec. 37). Localized lenses contain sufficient calcium carbonate to classify them as arenaceous limestones.

Fossils include *Crurithyris, Mesolobus,* and *Taonurus caudagalli* (Vanuxem). Specks of carbon and pieces of charcoal are mixed randomly in the calcareous sandstones and shales. Some scour and fill markings are exposed along the bedding planes.

Traced northwestward along the outcrop belt to northeastern Bates County, the sandstone is in most places even-bedded, but in limited areas it is massive and cross-bedded (Pl. 6, Sec. 24). It is fine-grained, subangular, micaceous, and weathers to a reddish-brown color. Several feet of conglomerate that consists of pea-sized particles of limestone, chert, and brachiopod fragments occur near the middle of the sandstone shale interval in the vicinity of Ballard in northeastern Bates County (Pl. 6, Sec. 43).

In the eastern part of Bates County the Lexington

coal occurs near the top of the formation. It is thin-bedded and commonly contains thin beds or lenses of dark shale (Pl. 6, Sec. 40).

The upper part of the Labette Formation contains two coal horizons approximately 4½ miles southwest of Butler (Pl. 6, Sec. 39). It has not been established with certainty which of these horizons develops into the Lexington coal bed of eastern Bates County and areas to the northeast of Bates County. Jewett (1941, p. 310), working in Kansas and Oklahoma, also recognized two coal horizons at this stratigraphic position. He tentatively correlated the upper bed with the Lexington coal bed at the type area in Lafayette County, Missouri. A coal bed occupying the same stratigraphic position as the lower unit was mined in former times along a tributary to Miami Creek about 2 miles northeast of Virginia in west-central Bates County. Both coal horizons appear to be absent in northwestern Vernon County (Pl. 6, Sec. 37).

The top several feet of the Labette Formation consist of shale, sandstone, calcareous shale, and arenaceous limestone. A thin conglomerate composed of limestone and shale particles occurs above the Lexington coal in east-central Bates County (Pl. 6, Sec. 41). These units grade into each other both horizontally and vertically over short distances.

The upper boundary of the Labette Formation has been placed at the base of the black fissile shale of the Anna Member. The Labette Formation lies unconformably on the Higginsville limestone in northwestern Vernon County.

PAWNEE FORMATION

The Lawrence Conference (in Moore, 1948, p. 2025) designated the Pawnee limestone a formation, and the four included units were given member status. Thus, the present classification subdivides the Pawnee Formation into four members that are, in ascending order, the Anna, Myrick Station, Mine Creek, and Coal City.

ANNA MEMBER

The Anna is the lowermost member of the Pawnee Formation and includes beds from the top of the Labette Formation to the base of the Myrick Station Member of the Pawnee Formation. The Anna Member is a very thin but persistent unit. It does not exceed 2 feet in thickness and in some places in western Missouri is less than 1 foot thick (Pl. 7).

The Anna Member is composed essentially of black fissile shale that in most exposures contains phosphatic concretions that vary in shape from irregular to spheroidal. Most of the phosphatic concretions are approximately one-half inch in diameter, with some spheroidal phosphatic concretions attaining diameters of 1¼ inches. A thin bed of soft shale that grades from black at the bottom to greenish-gray at the top overlies the black fissile shale in most areas, but in a few exposures this unit is absent and limestone of the Myrick Station Member lies directly on the black fissile shale (Pl. 7, Sec. 49). The only fossils found in the Anna Member are conodonts of the genus *Hindeodella*.

The upper boundary of the Anna Member is well marked by an abrupt lithologic change from shale to limestone of the Myrick Station Member. In this study the base of the Anna has been placed at the bottom of the black fissile shale that is one of the most persistent units of western Missouri. The top of the Lexington coal, which has been used as the base of the Anna Member in other parts of Missouri, was not selected, because the exact stratigraphic position of the Lexington coal has not been established in southwestern Bates and northwestern Vernon counties.

MYRICK STATION MEMBER

The Myrick Station Member includes the limestone beds between the Anna Shale Member and the Mine Creek Shale Member. The thickness of the Myrick Station is relatively constant throughout western Missouri. The average thickness is $3\frac{1}{2}$ feet (Pl. 7). The thickness rarely exceeds 5 feet, and at no outcrop was the Myrick Station Member observed to be less than 3 feet thick.

The member consists of finely crystalline, bluish-gray limestone that weathers brown. Bedding is thick and even. The main joint patterns are NW-SE and NE-SW. The fracture is angular or hackly. Fossils include poorly preserved recrystallized brachiopods and platy algae. Large, elongate fusulinids, some of which are 7 mm in length, occur at the top of the member (Pl. 7, Secs. 44, 50). The Myrick Station Member is overlain and underlain by shale. The gradation from the limestone of the Myrick Station Member to the shale is abrupt.

MINE CREEK MEMBER

The Mine Creek Member is predominantly shale and is underlain by the Myrick Station Limestone Member and overlain by the Coal City Limestone Member. The member varies considerably in thickness (Pl. 7). It is from 6 to 15 feet thick in Bates County and thins to about 1 foot in northwestern Vernon County.

The lower part of the Mine Creek Member is composed of dark-gray noncalcareous shale that grades upward into calcareous shale interbedded with limestone. The limestones are seldom more than a few inches thick and are dark-gray, argillaceous, thin-bedded, and jointed into blocks. They have the characteristic of weathering to a brownish-red color and locally may contain considerable fine sand (Pl. 7, Secs. 48, 49). The limestones and shales in the upper part of the member are abundantly fossiliferous and contain *Derbyia crassa* (Meek and Hayden), *Mesolobus mesolobus* (Norwood and Pratten), *Antiquatonia, Linoproductus,* and crinoid columnals. Fossils become more abundant near the top of the Mine Creek Member, and the shales and shaly limestones directly underlying the Coal City Member are a coquina of *Neochonetes granulifer* Owen, *Mesolobus mesolobus* (Norwood and Pratten), *Antiquatonia* sp., large crinoid columnals, and sparse bryozoans. The abundantly fossiliferous zone at the top of the Mine Creek Member persists throughout most of western Missouri and is an aid in identifying the member.

In northeastern Vernon County the Mine Creek Member thins considerably and is composed of a few inches of dark shale with a thin-bedded, dark-gray, limestone bed near the middle (Pl. 7, Sec. 44). The top few inches consist of dark-gray to black platy shale, with abundant *Lingula* and *Crurithyris* and sparse *Mesolobus*.

COAL CITY MEMBER

The Coal City Member is overlain by shale or clay belonging to the Bandera Formation and is underlain by fossiliferous shale and thin limestone beds of the Mine Creek Member of the Pawnee Formation. The Coal City Member is 13½ feet thick in northwestern Vernon County and thins to about 6 feet in northeastern Bates County (Pl. 7, Secs. 44, 50, respectively).

The member is composed predominantly of gray, wavy-bedded limestone. At most exposures the bedding is from 2 to 6 inches thick at the bottom and 1 inch or less at the top.

FIGURE 9.   Mine Creek Member of the Pawnee Formation, central Bates County, E½ SW¼ sec. 8, T. 40 N., R. 30 W. Pick has been placed just above one of several limestone beds that are present in the upper part of the member at most places. The limestone and shale beds near the top of the member (above pick handle) are abundantly fossiliferous with chonetid brachiopods.

In southwestern Bates County the upper 4 or 5 feet consists of thin, nodular-bedded, siliceous limestone that is overlain by about 1 foot of thick-bedded, blocky, siliceous limestone (Pl. 7, Sec. 45). The siliceous limestone weathers to angular pieces of white chert, which cover the slopes as a thick mantle of residuum. In northwestern Vernon County the top part of the member consists of interbedded shale and limestone, with some cherty limestone beds near the top (Pl. 7, Sec. 44).

FIGURE 10.   Coal City Limestone Member exposed in face of quarry, southwestern Bates County, SW¼ NW¼ sec. 24, T. 39 N., R. 33 W. The thick dark-gray limestone bed at top of quarry is very siliceous and weathers into blocks of white chert.

The upper surface of the Coal City Member commonly has an uneven or humpy appearance as a result of numerous colonies of *Chaetetes milleporaceous* Edwards and Haime. Other fossils include recrystallized brachiopods, *Fusulina* spp., and crinoid columnals.

The lower part of the Coal City Member in most places contains from a few inches to 1 foot of thick-bedded, blocky limestone with abundant *Mesolobus mesolobus decipiens* (Girty) and *Neochonetes granulifer* Owen. In many respects this limestone is more closely related lithologically and faunally to the thin limestone beds in the Mine Creek Member. When traced along the outcrop, this blocky limestone bed commonly becomes separated from the overlying wavy-bedded limestone by a shale lens, but in other areas it grades upward into wavy-bedded limestone and becomes lithologically similar to the Coal City limestone.

Therefore, the interval between the Coal City and Mine Creek members consists of interfingering limestone and shale lenses. For simplicity in mapping rock units the contact between the Mine Creek and Coal City members has been placed on top of the uppermost shale bed or lens.

BANDERA FORMATION

The Bandera Formation is underlain by the Coal City Member of the Pawnee Formation and overlain by the Amoret Member of the Altamont Formation.

The thickness of the Bandera Formation decreases irregularly along the outcrop belt from southwestern to northeastern Bates County (Pl. 8). The formation is over 50 feet thick in southwestern Bates County, but there is an abrupt decrease in thickness to 20 feet or less along the line of the Marais des Cygnes River between the towns of Amoret and Worland in western Bates County (Pl. 8, Secs. 54, 55). Northeastward from this area the formation decreases to 8 feet or less, but locally in central Bates County the formation increases in thickness to 35 feet or more (Pl. 8, Sec. 59).

The lithology of the Bandera Formation also varies considerably along the outcrop from southwestern to northeastern Bates County. South of a line, approximately the position of the Marais des Cygnes River, the formation is composed of shale, arenaceous shale, sandstone, coal, underclay, and minor amounts of limestone and conglomerate. The Mulberry coal lies near the base of the formation; it averages about 30 inches in thickness and has been extensively mined. It is underlain by 2 to 4 feet of underclay and in some places by an equal amount of light-gray calcareous shale. Overlying the Mulberry coal is approximately

30 feet of shale that locally contains large flattened septarian concretions composed of clay, with the cracks filled by coarsely crystalline limestone (Pl. 8, Sec. 53). The Bandera Quarry Sandstone Member lies near the top of the shale but in limited areas may be 30 or 40 feet thick and lie near the top of the Mulberry coal (Pl. 8, Sec. 52). The sandstone is commonly well bedded and is fine-grained and micaceous. In the highwall of some strip mines, the sandstone has been observed to grade laterally through arenaceous shale into shale. Locally it is absent. About 1 mile northeast of Hume the Bandera Quarry sandstone thickens abruptly to about 35 feet in a small synclinal structure (Pl. 8, Sec. 52). At this place it is lenticular and saturated with asphalt. Near the bottom it contains a thin conglomerate bed composed of particles of coal, shale, and limestone.

The Bandera Formation south of the Marais des Cygnes River contains few invertebrate fossils. Fragments of plants are abundant in the shale and arenaceous shale overlying the Mulberry coal.

North and northeast of the Marais des Cygnes River the Bandera Formation thins considerably. The Mulberry coal is irregular in thickness; decreases in thickness from 2 feet to a thin smut in a lateral distance of 25 feet have been observed (Pl. 8, Sec. 47). In some exposures the Mulberry coal contains thin lenses of dark shale (Pl. 8, Sec. 58). The shale overlying the Mulberry coal contains limestone of at least two different lithologies: (1) a thin, dark-gray concretionary limestone bed that locally overlies the Mulberry coal (Pl. 8, Secs. 57, 61), and (2) small limestone nodules that commonly occur near the top of the shale but locally may be present down to the Mulberry coal.

In most exposures north of the Marais des Cygnes

River, the Bandera Quarry sandstone is absent, but locally in central Bates County the Bandera Formation thickens to 35 feet or more (Pl. 8, Sec. 59). At this place the Bandera Quarry sandstone is represented near the top by thin shaly sandstone or isolated masses of sandstone. Several feet of dark-gray to black shale with numerous interbedded, thin coal lenses underlie the Mulberry coal. The shale contains abundant *Calamites* and in some places is almost completely composed of fragments of fossil plants. A thin conglomerate of very limited extent occurs near the bottom of the formation. The conglomerate is composed of pebble-sized particles of limestone and *Chaetetes* colonies. The shale and underclay below the Mulberry coal contain several lenses of siderite concretions. Most of the concretions are about 3 inches in diameter (Pl. 8, Secs. 59, 60).

A lower formational boundary is sharp in contrast to the upper boundary, which is gradational from shale into nodules of limestone mixed with clay. The upper boundary has been arbitrarily placed within the shale-nodular limestone sequence instead of at the bottom of it, mainly because it appears that some of the lowermost limestone nodules may grade laterally into a dark-gray concretionary limestone bed that is genetically related to the Mulberry coal.

ALTAMONT FORMATION

The Altamont Formation is composed of two relatively thin limestone beds separated by shale.

Apparently, the Altamont was considered to be one limestone unit until Jewett (1940a, p. 23) recognized that the formation consisted of two limestone units separated by a shale member. A few years earlier

Greene (1933, pp. 14-18), working independently in northern Bates and Cass counties, Missouri, recognized a similar sequence of two limestone units separated by shale. However, Greene believed that the sequence occurred in the Bandera shales and named the limestone beds the upper and lower Worland for exposures near Worland in western Bates County. As a result, Greene miscorrelated an overlying limestone, probably the Lenapah, with the Altamont.

In an attempt to alleviate the confusion that existed between Missouri and Kansas, a field conference was organized (Cline, 1941, p. 29), including R. C. Moore and J. M. Jewett of Kansas, L. M. Cline of Iowa, and F. C. Greene of Missouri. The participants in this conference traced the lower and upper Worland limestone units of Greene from the Missouri River, across southwestern Missouri, and into southeastern Kansas and showed them to be equivalent to the upper and lower Altamont limestones, respectively. It is not difficult to see why Greene considered the Altamont limestones to be part of the Bandera Formation, for just north of Worland the Bandera Formation thins to less than 10 feet in contrast to a thickness of 40 or 50 feet southwest of Worland (Pl. 8, Secs. 54, 55).

According to present classification the Altamont Formation consists of the strata between the Bandera and Nowata formations. It has been subdivided into three members that are, in ascending order, the Amoret, Lake Neosho, and Worland.

AMORET MEMBER

The Amoret Member includes the beds between the top of the Bandera Formation and the base of the Lake Neosho Member of the Altamont Formation. The Amoret Member varies in thickness from a few

FIGURE 11.   Altamont and Bandera formations exposed in highwall
of abandoned Mulberry coal strip mine near Worland, western
Bates County, NE¼ NW¼ sec. 7, T. 39 N., R. 33 W. The blocky
limestone bed at top of highwall is the Worland Member. The
Lake Neosho Member is represented by a thin dark-gray shale bed
below the Worland limestone. The Amoret Member, which is
composed of a few inches of nodular limestone, is barely visible
below the Lake Neosho Member. The Bandera Formation is
represented by approximately 40 feet of gray shale. Three miles
north of this location the Amoret thickens to include 5 feet or more
of bedded fossiliferous limestone, and the Bandera thins to 10 feet
or less.

inches to slightly over 8 feet. The member is thickest
at the type section in western Bates County and thins
irregularly northeast and southwest along the outcrop
belt (Pl. 8).

Lithologically, the Amoret Member consists of
nodules of limestone embedded in a clay matrix that
grades locally into bedded limestone that may be only
slightly nodular (Pl. 8, Secs. 54, 55, 56). The nodules
of limestone are in most places argillaceous; the percentage of clay increases with increasing nodularity.
At the type locality in western Bates County the
Amoret consists predominantly of limestone that is

wavy to nodular bedded (Pl. 8, Sec. 56). The bedded limestone of this area frequently has a granular appearance, which is the result of abundant specimens of *Osagia*. Brachiopod fossils include *Derbyia crassa* (Meek and Hayden), *Composita ovata* Mather, *Punctospirifer kentuckensis* (Shumard), *Antiquatonia portlockiana* (Norwood and Pratten), and several species of *Mesolobus*. An organ-pipe tabulate coral *Syringopora* spp. is found occasionally. The fossil content decreases with increasing nodularity.

The lithology of the Amoret Member at the type locality is not representative of the unit in most other areas of Bates County. Southwestward and northeastward along the outcrop belt from the type section in western Missouri the unit consists predominantly of sparsely fossiliferous nodules of limestone embedded in a greenish-gray clay. However, local thickening to abundantly fossiliferous bedded limestone does occur.

At most places the contact between the Amoret Member and the Bandera Formation is gradational from nodular limestone into clay or shale. At places where no lithologic break is visible between the limestone and shale, the formational contact is arbitrarily placed within the clay-nodular limestone sequence. The boundary between the Amoret Member and the Lake Neosho Member is somewhat less gradational and has been placed at the top of the limestone nodules or nodular limestone.

LAKE NEOSHO MEMBER

The Lake Neosho Member is overlain by the Worland Member and underlain by the Amoret Member. The Lake Neosho is a thin, persistent unit that averages about 3 feet in thickness. At no place was the member found to be more than 5 feet thick (Pl. 8).

The Lake Neosho Member consists of from 1 to 3 feet of dark-gray to black shale that grades vertically into light-gray calcareous shale. In most exposures the dark-gray to black shale contains large spheroidal phosphatic concretions, some of which are 1¼ inches in diameter. The center or core of most of the phosphatic concretions is hollow; however, some have formed around a crinoid columnal or other fossil fragment. The majority of the spheroidal phosphatic concretions are concentrically banded. The phosphatic concretions are present in most outcrops and aid in the identification of the unit. Nevertheless, spheroidal phosphatic concretions of equal size are present in the Anna Member, and other criteria must be included when identifying the Lake Neosho Member.

The top part of the Lake Neosho Member consists of a few inches of light-gray shale that contains *Mesolobus lioderma* Dunbar and Conrad and *Crurithyris planoconvexa* (Shumard). Specimens of very small *Mesolobus,* approximately one-sixteenth inch in length, are fairly abundant at this horizon. Wallace Howe of the Missouri Geological Survey (personal communication) has used these small brachiopods as an aid in the identification of the Altamont Formation. The base of the Lake Neosho Member has been placed on top of the limestone nodules, which are abundant in the Amoret Member in most areas.

WORLAND MEMBER

The Worland Member is underlain by the Lake Neosho Member of the Altamont Formation and is overlain by gray shales of the Nowata Formation. The Worland is from 3 to 6 feet thick (Pl. 8). It attains its greatest thickness in southwestern Bates County (Pl. 8, Sec. 51) and thins to approximately 3 feet in northeastern Bates County (Pl. 8, Sec. 50).

The Worland Member consists predominantly of limestone that is commonly light-gray, wavy-bedded, and jointed into large blocks or slabs that are several feet across. Main joint patterns are NW-SE and NE-SW. In the vicinity of Worland the lower surface of the member contains sinuous elevated structures about one-half inch in diameter. They are believed to be filled-in burrows or to have formed by differential compaction. These structures have the appearance of a piece of coiled rope and have been informally termed "ropy" structures. The upper surface of the member is irregular and pitted, probably as a result of differential solution by the processes of erosion. In places, colonies of *Chaetetes* give the upper surface a humpy appearance.

In southwestern Bates County the lower several inches of the Worland Member consists of thick-bedded limestone that is frequently jointed into blocks. A gray shale bed, 1 foot or less in thickness, separates this unit from the upper part, which is composed of 4 or 5 feet of thin, wavy-bedded limestone (Pl. 8, Sec. 51).

Common fossils are *Fusulina, Composita, Cleiothyridina, Crurithyris,* and small horn corals. Abundant platy algae frequently impart a minute wavy appearance to the outcrop surface.

In southwest-central Bates County, from Worland southeastward to the town of Sprague, a channel-filling, reddish-brown sandstone has locally replaced the Altamont Formation and the upper part of the Bandera Formation (Pl. 8, Sec. 54).

≋≋ Structural Geology

M ISSOURI IS DOMINATED by three major structural features, the Ozark uplift, the Missouri platform, and the Forest City basin (Fig. 12). The general dip of Paleozoic strata is northwestward off the Ozark uplift toward the Forest City basin. The attitude of the Paleozoic strata of northern and western Missouri has been modified by a series of northwestward-trending structures of regional extent. They were first described in detail by Hinds (1912, p. 8):

> All of these trend northwest-southeast and in places raise or lower the altitude of individual beds by 100 feet or more within short distances. The common mode of occurrence is a narrow well-marked syncline lying close to the southwest side of a rather narrow anticline. Beds commonly dip comparatively steeply to the southwest from the anticlines and more gently to the northeast. These folds, with the direction of their axes practically unchanged, can be traced for long distances across the State.

Hinds and Greene (1915, pl. XXIII) mapped in detail several broad, low, northwestward-trending anticlines and intervening synclines in the Pennsylvanian strata of Missouri. They recognized eight major anticlinal folds. These they named after the major towns through which the anticlinal axes trend. Approximate

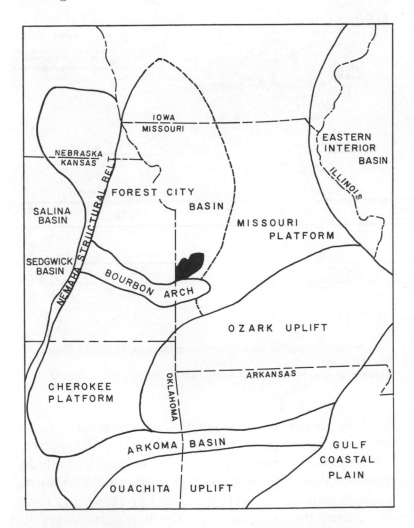

FIGURE 12.   Major Pennsylvanian structural features of the midcontinent area and their relationship to the project area (modified after Branson, 1962, p. 432). Project area shown by shading.

locations of the major anticlinal folds as recognized by
Hinds and Greene are shown in Figure 13.

FIGURE 13.   Structural map of the Pennsylvanian system of Missouri
showing northwest trending structures.
(Modified after Hinds and Greene, 1915, Plate XXIII)

More recent workers, particularly members of the
Missouri Geological Survey and Water Resources,
recognized that many structures that exist as folds in
younger Paleozoic rocks project downward into faults
or fault zones in subsurface rocks. The possibility then
exists that the areas between the major faults are in
reality fault blocks in subsurface rocks. Searight and
Searight (1961, p. 156) subdivided the state into several northwestward-trending structural segments:

The boundary between the Ozark dome and the Forest City basin is not a smooth, even incline, but it is an area broken by northwestward trending structural segments of considerable magnitude. Two of these are structurally depressed and two are elevated. For convenience, these are referred to as Tri-State Plateau, a depressed area in the southwestern part of the state; the west-central salient, a structurally elevated area in west-central Missouri; and east-central recess, a depressed area which separates the west-central salient from the Lincoln fold.

McCracken and McCracken (1965, map no. 6) recognized in subsurface studies of the Lower Ordovician a series of northwestward-trending structures of regional magnitude. Two of the structures that cross Bates and Vernon counties are believed by them to be probable faults.

W. C. Hayes (manuscript map on file at the Missouri Geological Survey, Rolla, Missouri) believes the Precambrian basement rocks of Missouri are dissected into tectonic blocks by a series of northwest-southeast and northeast-southwest-trending structural lineaments. The lineaments are considered by Hayes (personal communication) to be fault zones of undetermined width. Hayes recognized two structural lineaments in the Bates County area. The location and trend of the lineaments in the subsurface are similar in position to the axes of the Schell City-Rich Hill anticline and the Ladue-Freeman anticline.

The Bates County area lies on the southern margin of the Forest City basin (Fig. 12). Regionally, the Pennsylvanian strata of Bates County area dip northwestward toward the Forest City basin at 10 to 15 feet per mile. However, the attitude of the beds has been modified to a great extent by several northwestward-trending structures of regional extent (Pl. 1). The largest of these structures was named the Schell City-Rich Hill anticline by Hinds and Greene (1915, p. 206). This structure enters Bates County in the southeastern corner as a broad, low anticline and trends northwestward to near the Kansas-Missouri border in northwestern Bates County, where it plunges beneath younger sediments. The Schell City-Rich Hill anticline is asymmetrical. Dips of over 30° may be observed one mile west of Rich Hill, on the southwest limb, using the Higginsville limestone as a datum. The attitude of the northeastern limb is somewhat less steep. The closure is about 150 feet.

FIGURE 14.   Steeply inclined strata forming the southwestern limb of the Schell City-Rich Hill anticline, 1½ miles west of Rich Hill, SE¼ SW¼ sec. 1, T. 38 N., R. 32 W. Dip of Higginsville limestone recorded at small quarry is 35°S. 45°W. Photo taken looking south. Attitude and trend persist for 3 miles.

Detailed field mapping has disclosed that thick limestone beds, in particular the Higginsville, have been faulted along the steeply dipping southwestern limb of the Schell City-Rich Hill anticline. Structurally disturbed strata have been recognized also in rocks of Cherokee age. Faulting and considerable distortion of the shale and coal beds are present in strip-mining operations of the Mineral coal in northern Vernon County. The strip mines in which the faulting and distortion are the greatest are aligned with the southwestern limb of the Schell City-Rich Hill anticline. Hinds (1912, p. 76; Hinds and Greene, 1915, p. 461) reported steep dips in mines of the Rich Hill district. The elevation of the Rich Hill (Mineral) coal often varied as much as 150 feet in less than one-half mile. Dips steep enough to necessitate the use of special equipment for mine haulage were commonly encountered.

FIGURE 15.   Disturbed strata in west highwall of abandoned coal strip mine, north-central Vernon County, NE¼ sec. 34, T. 38 N., R. 31 W. Light-gray bed near center of photo is 4 feet thick and is composed of sandstone. It underlies the Croweburg coal in the Rich Hill area. The Mineral coal was mined. Structurally the section is on the southwestern flank of the Schell City-Rich Hill anticline. Dip is 30°S. 45°W. Photo taken looking west.

Faulting has not been observed in strata younger than the Higginsville limestone. Nevertheless, the difference in elevation of the Worland limestone at Worland and in exposures of this bed in areas along the Marais des Cygnes River, one mile to the north of Worland, indicates that rocks younger than the Higginsville were faulted.

When traced southwestward the faulting along the southwestern limb of the Schell City-Rich Hill anticline is aligned approximately with the Eldorado Springs fault zone in Cedar County (see Geologic Map of Missouri, 1961 edition). The Eldorado Springs fault zone dissects rocks of Pennsylvanian, Mississippian, and Ordovician age that have been exposed in the Ozark uplift.

A similar structure, the Ladue-Freeman anticline, crosses the northeastern part of Bates County. The area lying between these two structures appears to be a tilted fault block that dips at a slightly higher angle to the northwest than the fault block lying to the southwest of the Schell City-Rich Hill fault.

Several smaller folds, not more than a few square miles in area, are present on the southwestern limb of the Schell City-Rich Hill anticline (Pl. 1). They appear to plunge steeply northwestward, but at an angle of several degrees to the major structural axis of the Schell City-Rich Hill anticline. In other parts of the Bates County area, dips of 100 feet per mile have been observed; they appear to be related to structures of limited areal extent. These smaller folds of not more than a few square miles in areal extent are, in turn, wrinkled by still smaller minor structural features that vary in shape from asymmetrical folds to irregular dipping surfaces whose interrelationships could not be determined.

The Bourbon arch, an eastward extension of the Nemaha structural belt, lies directly south of the Bates County area and separates the Forest City basin from the Cherokee platform. The Bourbon arch is considered to have developed in late Mississippian time and to have persisted into early Pennsylvanian time (Lee, 1943; Jewett, 1945).

FIGURE 16.   Small asymmetrical fold in Atokan (?) strata, north side of drainage ditch, 1 ½ miles south of Prairie City, SW¼ NW¼ sec. 24, T. 38 N., R. 30 W. Axis of fold strikes N. 20° E. Structurally the fold is located near the axis of the Schell City-Rich Hill anticline.

On the basis of my limited reconnaissance work south of the project area in Vernon County, the upper Cherokee and lower Marmaton rocks do not seem to thin or to be otherwise appreciably different lithologically in the vicinity of the Bourbon arch than in Bates County. Therefore, it is assumed that the Bourbon arch was buried beneath lower Cherokee sediments and did not actually affect deposition of sediments of upper Cherokee and lower Marmaton age. The east-west trending nature of the structure is contrary to the major northwest-southeast trending structural features of the Bates County area. It is assumed that the eastward projection of the Bourbon arch has been faulted along the southwestern limb of the Schell City-Rich Hill anticline in areas where this structure cuts diagonally across the Bourbon arch.

The larger structural features of the Bates County area, particularly the Schell City-Rich Hill anticline, are considered to be the result of structural deformation resulting from diastrophic movements that acted intermittently over a long period of time. The structures of less than a few square miles may have resulted from one or more of the following processes: (1) solution and collapse of underlying carbonate rocks; (2) differential compaction; (3) slump; (4) irregularity of the original surface of deposition; or (5) true structural deformation resulting from diastrophism.

≋≋≋ Physical Variations in Cyclothems in the Bates County Area

CYCLOTHEMS of the Bates County area of western Missouri are characterized by variability. Although a few lithologic units of this area persist for miles without any noteworthy changes in physical characteristics, other units, when traced along the outcrop belt, were observed to vary considerably in physical characteristics. Some lithologic units that compose the cyclothem are facies changes of short lateral extent. This is particularly true of most of the sandstones, shales, and shaly limestones. Other units, particularly the purer limestones, black fissile shales, and coal beds, are very persistent. Nevertheless, many of the purer limestone and black fissile shale units were observed to be characterized by abrupt changes in thickness, lithology, and faunal characteristics in the Bates County area. Some coal beds, considered to be persistent in other areas, pinch out in this area, and their stratigraphic positions could be determined only by a study of the associated strata.

Noteworthy examples of stratigraphic units that vary in thickness or lithology are as follows: (1) The Verdigris (Ardmore) limestone, which consists of two beds of well-bedded limestone separated by shale in northwestern Vernon and southwestern Bates County, becomes a single bed of nodular-to-bedded limestone in the vicinity of the Schell City-Rich Hill anticline in south-central Bates and north-central Vernon County. Although it thins somewhat irregularly, it persists along the outcrop belt as a nodular limestone bed to at least western Henry County. In central Henry County, a limestone facies similar to that in southwestern Bates and northwestern Vernon County is once again encountered. (2) The black fissile shale underlying the Verdigris limestone is over 2 feet thick in the southwestern part of the Bates County area. When traced northeastward to the vicinity of the southwestern flank of the Schell City-Rich Hill anticline, the black fissile shale thins to a few inches of soft greenish-gray shale but retains the phosphatic concretions characteristic of the black fissile facies. The greenish-gray shale facies persists to western Henry County. According to Searight and Searight (1961, p. 161) the black fissile shale grades into greenish-gray shale toward mildly positive areas. (3) The black fissile shale of the Excello Formation thins and loses its fissility in some areas along the axis of the Schell City-Rich Hill anticline. (4) The Mulky coal, which is from several inches to 3 feet thick in Henry and Vernon counties, is represented by a thin coal smut throughout most of Bates County. (5) The Wheeler coal bed and associated strata are not present throughout much of southeastern and south-central Bates County. (6) This is also true of the coal bed or beds in the lower part of the Mulky Formation, which could not be traced southwestward from northeastern Bates County. (7) The Bevier coal was not recognized .southwestward from western Henry County. (8) The Amoret limestone changes abruptly from an unfossiliferous, nodular limestone to an abundantly fossiliferous, bedded limestone along the axis of the Schell City-Rich Hill anticline in western Bates County, between the towns of Worland and Amoret.

The Mineral coal, which attains a thickness of over 6 feet in the Rich Hill coal field, is the thickest coal bed in Missouri. The Rich Hill field is structurally a northwest-southeast-trending elongate basin, located along the southwestern limb of the Schell City-Rich Hill anticline.

The variation in thickness of the Verdigris and Bandera formations is particularly interesting when they are traced along the outcrop belt from Kansas to the vicinity of the southwestern limb of the Schell City-Rich Hill anticline. The thickness of the Verdigris Formation increases from 25 to 45 feet, while the Bandera Formation thins from over 50 feet to less than 20 feet (Pls. 3 and 8). It is doubtful if the thinning of the Bandera Formation, which is predominantly shale, has resulted entirely from the compaction of sediments. Lithologically, no difference has been observed in the character of the shales in areas to the southwest of the Schell City-Rich Hill anticline, where over 50 feet of shale has been recorded in numerous sections, and in areas to the northeast of this structure, where the total formational thickness is frequently less than 10 feet. Consequently, the thinning of the Bandera Formation in areas northeast of the Schell City-Rich Hill anticline is considered to have resulted predominantly from nondeposition.

In other areas of Missouri, preliminary field mapping has disclosed considerable variation in the physical characteristics of the Higginsville limestone and the Lagonda shale in the vicinity of the Centerview-

Kansas City and the College Mound-Bucklin anticlines, respectively. The Higginsville limestone, when traced northeastward along the outcrop belt to the vicinity of the Centerview-Kansas City anticline in north-central Johnson County, varies in a lateral distance of a few miles from a crinkly-bedded pure limestone approximately 15 feet thick to a nodular, argillaceous limestone unit less than 4 feet thick.

The Lagonda Formation, which consists predominantly of shale and sandstone, is over 50 feet thick in northeastern Chariton County and thins abruptly to less than 15 feet when traced northeastward along the outcrop belt to the vicinity of the College Mount-Bucklin anticline in southern Macon County. Still other examples can be cited.

≋ Relationship of Structural Movement to Stratigraphy

T HE PERSISTENCE and physical characteristics of a particular stratigraphic unit of a cyclothem in limited areas do not appear to be completely controlled by the mechanism or mechanisms responsible for the extent of the transgressions or regressions of Pennsylvanian seas over wide areas. The lithology and thickness of some of the units appear to have been influenced by irregularities on the depositional surface at the time of deposition, because most variations in lithology and thickness are associated with mappable structures of local or regional extent. Irregularities on a depositional surface may originate by one or a combination of the following processes: (1) solution and collapse of the underlying Mississippian or older carbonate rocks; (2) differential compaction of the underlying, earlier-deposited, Pennsylvanian sediments; (3) irregularities in the original surface of deposition; (4) structural deformations synchronous with subsidence of the major basins, such as the Forest City.

It is unlikely that processes other than structural deformation were the dominant causes for the variation in physical characteristics of the lithologic units in the Bates County area. Process designated (1) would be operative only over the limited areas and could not affect an area of tens of square miles, such as the Bates County area. Moreover, many of the depositional irregularities recognizable in the lithologic units in the Bates County area appear to have resulted from positive movement of the original surface of deposition and not from downward movement, as would occur in areas of solution and collapse of the underlying strata (Figs. 17a and 17b).

Process designated (2) is unlikely, because the thickness of individual coal beds is not related to the lithology of the underlying rocks. Coal beds of minable thickness are in many places almost directly underlain by relatively noncompactible beds of limestone and sandstone (Pls. 4 and 8).

Process designated (3) requires a mechanism capable of producing a series of irregularities on the original surface of deposition intermittently during the Pennsylvanian Period and in the same geographic location. The most likely mechanism to account for this phenomenon in the Bates County area is structural deformation that acted intermittently during the time interval corresponding to the deposition of the upper Cherokee and the lower Marmaton groups. Structural deformation must also be considered as the most likely possibility for the following reason:

Lithologic and paleontologic features of Pennsylvanian cyclical deposits indicate that deposition took place in shallow marine water, brackish water, or fresh water near sea level. In the Bates County area a maximum of 800 feet of Pennsylvanian strata that were deposited under these conditions overlie Mississippian rocks. It must be concluded that several hundred feet of subsidence occurred, in order to account for the de-

FIGURE 17a.   Croweburg coal lying horizontally on disturbed
sandstone and shale beds which pinch out near top of small fold,
east-central Bates County, NW¼ NE¼ sec. 22, T. 40 N., R. 29 W.

FIGURE 17b.   Close-up of faulting on west limb of structure shown
in Figure 17a. Displacement of the Croweburg coal is 2 feet. It is
assumed that the sandstone beds pinch out through
nondeposition near the crest of a small fold. At some later time
(after the Croweburg coal was deposited horizontally on the
underlying strata) the southwestern flank of the fold was faulted by
renewed movement.

position of this thickness of strata. It is reasonable to assume that the regional subsidence in the Bates County area was north or northwestward into the Forest City basin and was accompanied by minor undulations that could have affected sedimentation. Variations in the thickness, lithology, and faunal characteristics of the units composing the cyclic deposits of western Missouri may be the result of these crustal disturbances.

According to Weller (1957, pp. 326-328), the Pennsylvanian, more than any previous Paleozoic Period, was a time of crustal instability. During the Pennsylvanian Period the central and eastern parts of the United States existed as a somewhat irregular depression, bounded by a low upland to the north, higher areas to the east and south, and irregular, lower mountains to the west. This interior region was interrupted by the Cincinnati arch with its northern branches; the Ozark dome; the Michigan basin; the Illinois-Indiana-Western Kentucky basin; and the Forest City basin. Important but less pronounced structures, such as the Nemaha structural belt of eastern Kansas and southeastern Nebraska and the LaSalle anticline and DuQuoin axis in Illinois, came into existence. Finally, many minor structures either were accentuated or made their appearance (Weller, 1957, p. 328).

In the basins, subsidence predominated, interrupted by periodic, brief intervals of minor uplift that resulted in characteristic cycles of deposition. Minor irregularities—local unsynchronized movements of either uplift or depression—may account for uncommon irregularities in the stratigraphic sequences that do not entirely fit into the general cyclical pattern (Weller, 1957, p. 363). Lithologic units that crop out in the Bates County area several hundred feet above sea level are found at a depth of several hundred feet below sea level in the Forest City basin. Thus, hundreds or even possibly thousands of feet of subsidence occurred in the Forest City basin during the Pennsylvanian Period.

It is reasonable to assume that subsidence of the Bates County area in accordance with the Forest City basin was accompanied by diastrophic movements of the area adjacent to the Schell City-Rich Hill fault. Undifferentiated movements of structural blocks would alter considerably the deposition of sediments that was taking place in the near-shore type of environment that predominated in the Pennsylvanian Period. The proximity of the variations in physical characteristics of the units to the southwestern limb of the Schell City-Rich Hill anticline is more than fortuitous. Movements associated with the structure are considered the most likely mechanism to account for the variations.

The area between the Schell City-Rich Hill and the Ladue-Freeman anticlines, which in reality may be a structural block bounded by deep-seated faults, appears to have been active intermittently throughout the time interval corresponding to deposition of the upper Cherokee and lower Marmaton groups. Periods of greatest activity appear to coincide with intervals containing predominantly nonmarine units and include the Fleming, Croweburg, Verdigris, Bevier-Lagonda, Mulky, Labette, Bandera, and the lower part of the Altamont Formation. Probably the structure was in a state of quiescence during the time corresponding to the deposition of the thick marine limestone of Pawnee Formation and Fort Scott Subgroup, as only slight lithologic variations are observed in these units.

In the southern part of the Bates County area, during deposition of the Verdigris Formation, the structural block first subsided more rapidly than the surrounding area. This movement resulted in an expansion of the lower part of the Verdigris Formation from the top of the Croweburg coal to the base of the black fissile shale underlying the Verdigris limestone. Later, the block subsided less rapidly than the surrounding area, with the result that the strata composing the upper part of the Verdigris Formation are reduced in thickness and contain features characteristic of first a shoal area and then of a slightly emergent region.

The relationship of structure to the physical variations in the Verdigris Formation is shown in Figure 18.

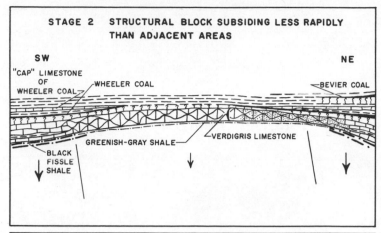

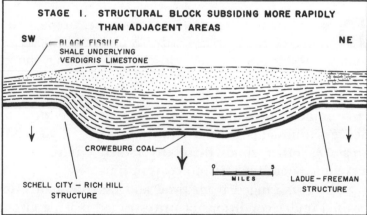

FIGURE 18. Diagrammatic cross-sections showing in chronological sequence the interrelationships between structure and sedimentation in the area from southwestern Bates County to west-central Henry County during the time interval corresponding to deposition of the Verdigris Formation. Movement of the structural block between the Schell City-Rich Hill and the Ladue-Freeman structures may have been the cause for the physical variations observed in the Verdigris Formation.

≋≋ Paleogeography

DURING THE PENNSYLVANIAN PERIOD, seas repeatedly transgressed the Bates County area and flooded a somewhat uneven surface close to a shifting shore. The water was salty, brackish, or fresh and everywhere shallow but of unequal depth. Numerous depositional environments prevailed in the area. Sometimes diversified environments existed simultaneously in adjacent areas. At other times these diversified environments were separated by short intervals of time.

Sediments ranged from lime and detrital muds deposited under conditions of unrestricted marine circulation to black organic muds laid down in a restricted lagoonal or tidal-flat environment. The withdrawal of the shallow seas into the deeper parts of the Forest City and Arkoma basins caused streams to extend their courses over the newly emergent region. Sand, silt, and mud were deposited in the form of deltaic, floodplain or channel-fill deposits.

The record of these gradually shifting environments has been preserved in the stratigraphic sequence as facies of diverse lithology. According to most students of the Pennsylvanian Period, the major direction of sediment transportation is generally believed to have been from the north or northeast (Wheeler and Murray, 1957, p. 1990). Nevertheless, the Nemaha structural belt of Kansas and the Ouachita Mountain region of Oklahoma and Arkansas could have contributed considerable quantities of sediment to the area. If direction of sediment transport was from the north or northeast, it would have been necessary for some sedimentary units deposited in a nonmarine environment to have had the sediment that composes them transported across areas where marine or lagoonal sedimentation was in progress (Fig. 20a).

The Ozark dome is considered to have been a minor source area, which at times was inundated by marine transgressions. Gradation from a marine to nonmarine environment upon approaching the Ozark dome is indicated by sedimentological studies of some of the units. It is assumed that during most of the Pennsylvanian Period the Ozark dome was a low emergent area and that marine waters inundated the region around it.

Numerous northwest-southeast-trending structures extended from the Ozark dome into the Forest City basin. The Schell City-Rich Hill anticline was the most prominent of these structures in the Bates County area. Intermittent movement along the structure influenced the depositional environment throughout much of the area.

The paleogeography of the Bates County area during deposition of the Verdigris Formation, the lower part of the Bevier Formation and parts of the Bandera, and the Altamont Formation is reconstructed on paleosedimentological maps arranged in chronological sequence (Figs. 19 and 20). The environments of deposition corresponding to each map are discussed in detail in the following paragraphs.

Following the deposition of the Croweburg coal, a nonmarine to brackish water environment prevailed in the Bates County area. Detrital mud, sand, and silt were being deposited to form deltas or tidal flats (Fig. 19a). Thickest accumulations were in the Rich Hill area in the southern part of the Bates County area (Pl. 3).

The tidal flat or deltaic environment was followed by marine transgression, and a restricted lagoonal environment came into existence. In the western half of the area, black organic muds containing a high percentage of phosphate were formed in an environment of restricted marine circulation. The eastern half of the region at this time was aerated sufficiently to cause oxidation of the organic matter, with the result that greenish-gray phosphatic muds were deposited. It is postulated that the eastern half of the Bates County area was a shoal area and that waves agitated the bottom sediments. The restricted lagoonal environment was gradually replaced by an environment of less restricted circulation (Fig. 19b). A eustatic rise in sea level or a lowering of the land surface by epeirogenetic movements allowed marine waters to enter the area, which had previously been a lagoon. Lime muds and detrital muds were deposited in this newly formed marine or lagoonal environment in the western part of the area, while near-shore, brackish water deposits of sand, mud, and lime were being deposited in the northeast. Throughout most of the central part of the area the restricted environment was gradually replaced by marine encroachment that was accompanied by deposition of lime muds. Progressive marine transgression continued, and beds of lime and detrital mud were deposited in the western third of the area in an environment of normal marine circulation, while in the eastern two-thirds, the shoal-type environment, which came into existence during deposition of the underlying phosphatic shales, persisted with deposition of lime muds (Fig. 19c). A facies change, from a well-bedded limestone in the west to nodular limestone in the northeast, infers that waves agitated the bottom sediments in the eastern part of the area. The northeastern part of the area continued to receive considerable amounts of sand derived from some unknown source area. Following deposition of the Verdigris limestone, marine waters withdrew from the area. It is a likely supposition that deltaic sandstones and shales were deposited during the nonmarine portion of the cyclothem but were removed subsequent to the development of the underclay of the Wheeler coal.

Renewed marine transgressions, originating in the basin areas to the southwest or northwest, resulted in the gradual encroachment of the seas over the region. For a time coal swamps flourished in a fresh or brackish water environment in the western part of the area, but this environment was replaced gradually by the development of a near-shore marine environment (Fig. 19d). As the seas gradually transgressed over the region, marine limes and muds were deposited over beds of peat in the western part of the area. These now compose the fossiliferous shale and argillaceous limestone overlying the Wheeler coal. Minor transgressions resulted in the interlamination of thin beds of marine lime and shale with peat in the brackish or fresh water environment that existed to the east. These deposits are now represented in the section as an alternating sequence of thin laminae of coal and fossiliferous marine limestone, which are present at some places in the upper inch or less of the Wheeler coal. Most of the central part of the area was elevated slightly, so that organic matter was oxidized, with the result that the Wheeler coal is absent in this area. The thickness pattern of the Verdigris Formation is shown in Figure 19e.

During the time interval corresponding to deposition of the upper part of the Bandera Formation, thick beds of sand and mud were accumulating to form deltas and tidal flats in the southwestern part of the area (Fig. 20a). Meanwhile, the remainder of the region was receiving accumulations of lime mud and detrital mud in a lagoonal environment. Isolated lenses of sand and detrital mud occur in the middle of the lagoonal environment. It is postulated that these isolated areas were small, structurally controlled basins that were subsiding more rapidly than the surrounding area. Detrital mud and sand were the major sediments carried into them by currents, lime mud being a minor constituent. (I have studied small bodies of sand and silt in an area of black mud deposition in the Mississippi Sound off the southern coast of the United States. Currents appear to have winnowed the coarser fraction from the black muds.)

Progressive marine transgression continued, and beds of lime and detrital mud were deposited in areas where the bottom sediments were not agitated to an appreciable degree (Fig. 20b). Throughout the major part of the Bates County area shoaling conditions prevailed, and lime muds were shaped into nodules by wave agitation of bottom sediments. In the southwestern part of the Bates County area, detrital mud and sand continued to be deposited in the form of deltas or on tidal flats.

Gradation from a lagoonal environment in the

northeast to a predominantly nonmarine deltaic environment in the southwest implies that sediment was derived from a source area that lay to the southwest, west, or northwest. However, until studies of a more regional nature can be completed, this conclusion is tentative and premature. The combined thickness of the Bandera and Altamont formations is shown in Figure 20c.

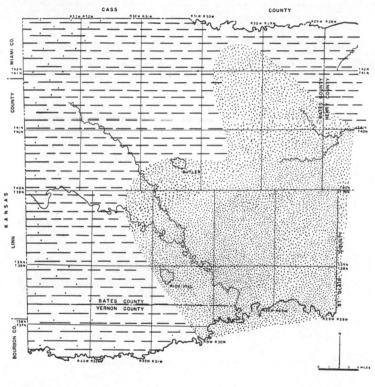

Lagoonal environment, restricted circulation—deposition of black mud and organic phosphate

Delta to tidal flat environment—deposition of detrital mud

Delta to tidal flat environment—deposition of sand and silt

FIGURE 19.   Paleosedimentological maps, Verdigris Formation (a) Sandstone, gray shale, and phosphatic shale interval beneath Verdigris limestone.

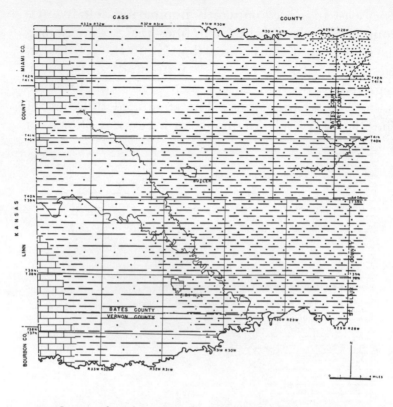

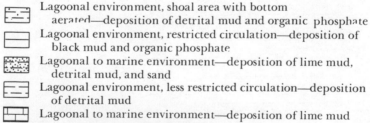

Lagoonal environment, shoal area with bottom
aerated—deposition of detrital mud and organic phosphate

Lagoonal environment, restricted circulation—deposition of
black mud and organic phosphate

Lagoonal to marine environment—deposition of lime mud,
detrital mud, and sand

Lagoonal environment, less restricted circulation—deposition
of detrital mud

Lagoonal to marine environment—deposition of lime mud

FIGURE 19.   (continued), (b) Phosphatic shale beneath the
Verdigris limestone and the lower part of the Verdigris limestone.

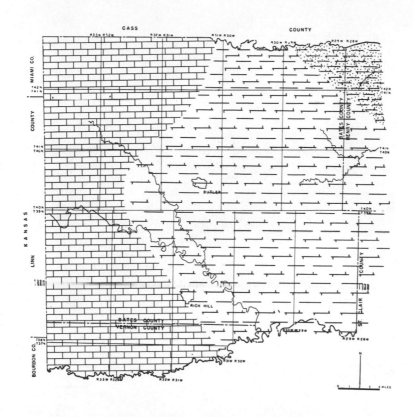

    Marine environment—deposition of lime mud

    Marine to lagoonal environment—deposition lime mud and
        detrital mud

    Marine to lagoonal environment—deposition of lime mud
        and detrital mud and sand

FIGURE 19.   (continued), (c) Verdigris limestone.

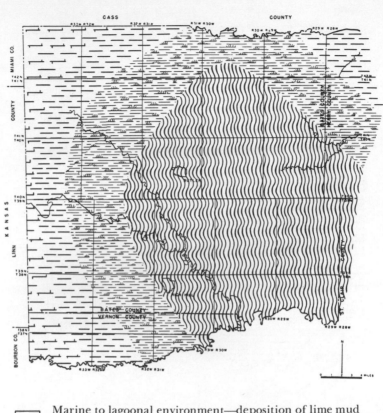

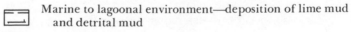

Marine to lagoonal environment—deposition of lime mud and detrital mud

Lagoonal environment—deposition of detrital mud

Coal swamp environment—deposition of peat

Low elevated aerated region or mud flat

FIGURE 19.   (continued), (d) Wheeler coal, underclay, and lower part of Bevier Formation.

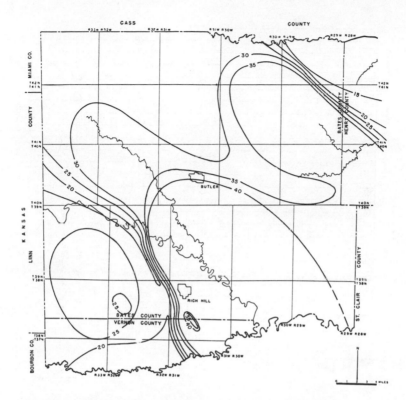

F<small>IGURE</small> 19.   (continued), (e) Isopach map of the Verdigris
Formation.

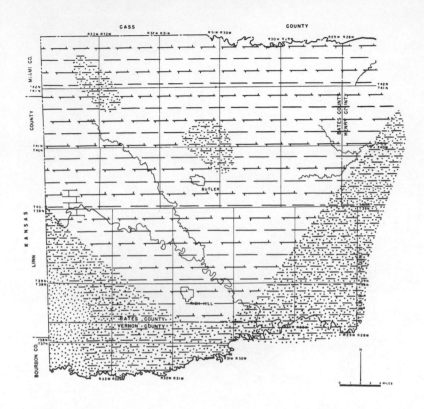

 Lagoonal environment—deposition of lime mud

Lagoonal to tidal flat environment—deposition of detrital mud and lime mud

Delta to tidal flat environment—deposition of detrital mud and sand

Delta to tidal flat environment—deposition of detrital sand

FIGURE 20.   Paleosedimentological maps, upper part of the Bandera Formation and lower part of the Altamont Formation (a) Interval beneath Amoret Limestone Member of the Altamont Formation.

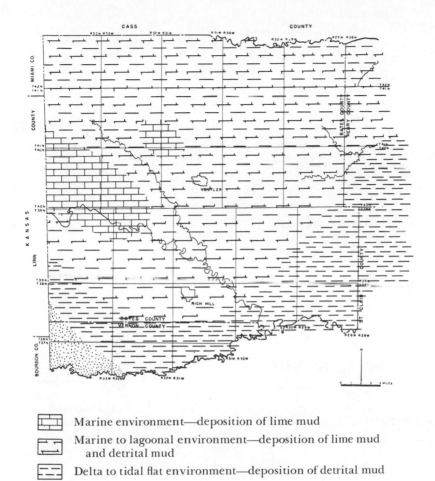

Marine environment—deposition of lime mud

Marine to lagoonal environment—deposition of lime mud and detrital mud

Delta to tidal flat environment—deposition of detrital mud

Delta to tidal flat environment—deposition of detrital sand

FIGURE 20.　(continued), (b) Upper part of the Bandera Formation and lower part of the Amoret Limestone Member.

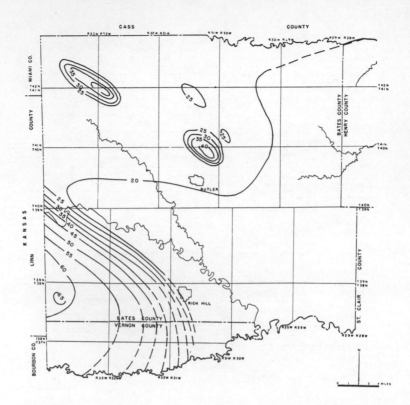

FIGURE 20.   (continued), (c) Isopach map of the Bandera and Altamont formations.

≋≋≋ Conclusions

THE VARIATIONS in physical characteristics of many thin lithologic units in western Missouri are best explained by structural deformation of the surface of deposition contemporaneous with sedimentation. Regional mapping has disclosed that the greatest amount of physical variation of the units occurs in the vicinity of northwest-trending, broad, low anticlines of regional extent. A roughly parallel series of these structures has been mapped in the Pennsylvanian strata of northern and western Missouri. The anticlines approximate the position of deep-seated faults when projected into the subsurface.

The most prominent of these structures in western Missouri is the Schell City-Rich Hill anticline. The southwestern limb of this structure is truncated by a high-angle fault that has been recognized in subsurface mapping of early Paleozoic and Precambrian rocks. Significant variations in physical characteristics of individual units have been observed in the upper Cherokee and lower Marmaton groups when mapped along the outcrop belt from the Kansas-Missouri boundary to the southwestern limb of the structure. Structural movements of the areas adjacent to the faulted southwestern limb of Schell City-Rich Hill anticline do not appear to have been synchronized in magnitude or direction, and as a consequence the resulting cyclical sequences do not show the same degree or type of lithologic variation.

Bibliography

BRANSON, C. C., 1962, The Pennsylvanian System in the United States, a symposium; Am. Assoc. Petroleum Geologists, 508 p.

BROADHEAD, G. C., 1874, Geology of Bates County: Missouri Geol. Survey, Rept. of 1873-1874, p. 155-178.

———— & C. J. NORWOOD, 1874, Geology of Vernon County: Missouri Geol. Survey, Rept. of 1873-1874, p. 119-154.

CLAIR, J. R., 1943, The oil and gas resources of Cass and Jackson Counties, Missouri: Missouri Geol. Survey, 2nd ser., v. 27, 208 p.

CLINE, L. M., 1941, Traverse of upper Des Moines and lower Missouri Series from Jackson County, Missouri to Appanoose County, Iowa: Am. Assoc. Petroleum Geologists Bull., v. 25, no. 1, p. 23-75.

———— & F. C. GREENE, 1950, A stratigraphic study of the upper Marmaton and lowermost Pleasanton Groups, Pennsylvanian, of Missouri: Missouri Geol. Survey, Rept. of Inv. 12, 74 p.

DAKE, C. L., & JOSIAH BRIDGE, 1932, Buried and resurrected hills of the central Ozarks: Am. Assoc. Petroleum Geologists Bull., v. 16, p. 629-652.

DAPPLES, E. C., W. C. KRUMBEIN, & S. L. SLOSS, 1948, Tectonic control of lithologic associations: Am. Assoc. Petroleum Geologists Bull., v. 32, no. 10, p. 1924-1947.

DUNBAR, C. O., & G. E. CONDRA, 1932, Brachiopoda of the Pennsylvanian System in Nebraska: Nebraska Geol. Survey, Bull. 5, 2nd ser., 377 p.

ELIAS, M. K., 1937, Depth of deposition of the Big Blue (Late Paleozoic) sediments in Kansas: Geol. Soc. Am. Bull., v. 48, p. 403-432.

FRIEDMAN, S. A., 1960, Channel-fill sandstones in the middle Pennsylvanian rocks of Indiana: Indiana Dept. of Conservation, Rept. of Prog. 23, 59 p.

GORDON, C. H., 1893, Report on the Bevier sheet; including portions of Macon, Randolph, and Chariton Counties: Missouri Geol. Survey, vol. 9, Sheet Report 2, 1st ser., 85 p.

GRAY, H. G., 1961, Determinants of cyclic sedimentation in Pottsville rocks near Dundee, Ohio: Ohio Jour. Sci., v. 61, no. 6, p. 353-366.

GREENE, F. C., & W. F. POND, 1926, The geology of Vernon County: Missouri Bur. Geol. and Mines, 2nd ser., v. 19, 152 p.

————, 1933, Oil and gas pools of western Missouri: Missouri Bur. Geol. and Mines, 57th Bienn. Rept. State Geologist, 1931-1932, app. 2, 68 p.

———— & W. V. SEARIGHT, 1949, Revision of the classification of the post-Cherokee Pennsylvanian beds of Missouri: Missouri Geol. Survey, Rept. Inv. 11, 21 p.

HARBOUGH, JOHN, 1962, Geoeconomics of the Pennsylvanian marine banks in southeast Kansas: Kansas Geol. Soc. and State Geol. Survey of Kansas, Guidebook, 27th Annual Field Conf., 160 p.

HAYES, M. D., 1963, Petrology of Krebs Subgroup (Pennsylvanian, Desmoinesian) of western Missouri: Am. Assoc. Petroleum Geologists Bull., v. 47, no. 8, p. 1537-1551.

HEDBERG, H. D., 1936, Gravitational compaction of clays and shales: Am. Jour. Sci., ser. 5, v. 31, p. 241-287.

HINDS, HENRY, 1912, Coal deposits of Missouri: Missouri Bur. Geol. and Mines, 2nd ser., v. 11, 503 p.

———— & F. C. GREENE, 1915, The stratigraphy of the Pennsylvanian Series in Missouri: Missouri Bur. Geol. and Mines, 2nd ser., v. 13, 407 p.

HOARE, R. D., 1961, Desmoinesian brachiopoda and mollusca from southwest Missouri: Univ. of Missouri Studies, v. XXXVI, 263 p.

HOVER, F. B., 1958, Geology of the east half of the Johnston and Creighton Quadrangles, Henry County, Missouri: unpublished Master's thesis, Univ. of Missouri, Columbia, Missouri, 92 p.

HOWARD, L. W., & W. H. SCHOEWE, 1965, The Englevale channel sandstone: Trans. Kansas Academy of Science, v. 68, no. 1, p. 88-106.

HOWE, W. B., 1953, Upper Marmaton strata in western and northwestern Missouri: Missouri Geol. Survey, Rept. Inv. 9, 29 p.

————, 1956, Stratigraphy of the pre-Marmaton Desmoinesian (Cherokee) rocks in southeastern Kansas: Kansas State Geol. Survey, Bull. 123, 132 p.

HUDDLE, J. W., & S. H. PATTERSON, 1962, Origin of Pennsylvanian underclay and related seat rocks: Geol. Soc. of America Bull., v. 72, no. 11, p. 1643-1660.

HUFFMAN, G. G., 1959, Pre-Desmoinesian isopachous and paleogeologic studies in central Mid-Continent region: Am. Assoc. Petroleum Geologist Bull., v. 43, no. 11, p. 2541-2574.

JEFFERIES, N. W., 1958, Stratigraphy of the lower Marmaton rocks of Missouri: unpublished Doctoral dissertation, Univ. of Missouri, Columbia, Missouri, 331 p.

JEWETT, J. M., 1940*a*, Oil and gas in Linn County, Kansas: Kansas Geol. Survey, Bull. 30, 29 p.

———, 1941, Classification of the Marmaton group: Kansas Geol. Survey, Bull. 38, pt. 11, p. 285-344.

———, 1945, Stratigraphy of the Marmaton Group, Pennsylvanian, in Kansas: Kansas Geol. Survey, Bull. 58, 148 p.

———, 1951, Geologic structures in Kansas: Kansas Geol. Survey, Bull. 90, pt. 6, 66 p.

KRUMBEIN, W. C., 1948, Lithofacies maps and regional sedimentary stratigraphic analysis: Am. Assoc. Petroleum Geologists Bull., v. 32, no. 10, p. 1909-1923.

———, 1952, Principles of facies map interpretation: Jour. Sedimentary Petrology, v. 22, no. 4, p. 200-211.

——— & L. L. SLOSS, 1963, Stratigraphy and sedimentation: W. H. Freeman Pub. Co., 2nd ed., 660 p.

LANDES, K. K., 1959, Petroleum geology: John Wiley & Sons, Inc., p. 282-291.

LEE, WALLACE, 1943, Stratigraphy and structural development of the Forest City basin in Kansas: Kansas Geol. Survey Bull. 51, 142 p.

——— and others, 1946, Structural development of the Forest City basin of Missouri, Kansas, Iowa, and Nebraska: U.S. Geol. Survey, Oil and Gas Inv., Preliminary Map 48.

LOWRY, W. D., 1957, Implications of gentle Ordovician folding in western Virginia: Am. Assoc. Petroleum Geologists Bull., v. 41, no. 4, p. 643-655.

MANOS, CONSTANTINE, 1967, Depositional environments of the Spar land cyclothem (Pennsylvanian), Illinois and Forest City basins: Am. Assoc. Petroleum Geologists Bull., v. 51, no. 9, p. 1843-1861.

MEHL, M. G., 1920, The influence of the differential compression of sediments on the attitude of bedded rocks (abst.): Science, v. 51, p. 520.

MISSOURI GEOL. SURVEY: Geologic map of Missouri; 1961.

MOORE, R. C., 1935, Late Paleozoic crustal movements of Europe and North America: Am. Assoc. Petroleum Geologists Bull., v. 19, no. 19, p. 1253-1307.

———, 1948, Classification of the Pennsylvanian rocks in Iowa, Kansas, Missouri, Nebraska, and Oklahoma: Am. Assoc. Petroleum Geologists Bull., v. 32, no. 11, p. 2011-2040.

———, 1949, Divisions of the Pennsylvanian System in Kansas: Kansas Geol. Survey, Bull. 83, 203 p.

——— and others, 1951, The Kansas rock column: Kansas Geol. Survey, Bull. 89, 132 p.

———, 1957, Geological understanding of cyclic sedimentation represented by Pennsylvanian and Permian rocks of northern Midcontinent region: Guidebook, 21st Annual Field Conf., Kansas Geol. Soc., p. 77-84.

MUELLER, J. C., & H. R. WANLESS, 1957, Differential compaction of Pennsylvania sediments in relation to sand-shale ratios, Jefferson County, Illinois: Jour. Sedimentary Petrology, v. 27, p. 80-88.

Muir-Wood, Helen, & G. A. Cooper, 1960, Morphology, classification and life habits of the Productoidea (Brachiopoda): Geol.
Soc. of America, Mem. 81, 447 p.

Nevin, C. M., & R. E. Sherrill, 1929, Studies in differential compaction: Am. Assoc. Petroleum Geologists Bull., v. 13, no. 1, p. 1-22.

Payton, C. E., & L. A. Thomas, 1959, The petrology of some
Pennsylvanian black "shales": Jour. Sedimentary Petrology, v. 29,
no. 2, p. 172-177.

Potter, P. E., 1962, Shape and distribution patterns of Pennsylvanian sand bodies in Illinois: Illinois Geol. Survey, Circ. 339, 36 p.

———, 1963, Late Paleozoic sandstones of the Illinois Basin: Illinois
Geol. Survey, Rept. Inv. 217, 92 p.

Rubey, W. W., & N. W. Bass, 1925, The geology of Russell County,
Kansas: Kansas Geol. Survey, Bull. 10, pt. 1, p. 72-85.

Rusnak, G. A., 1957, A fabric and petrologic study of the Pleasant-
view sandstone: Jour. of Sedimentary Petrology, v. 27, no. 1, p. 41-
45.

Salas, J. A. J., & J. M. Serratosa, 1953, Compressibility of clays:
Proc. 3d Intern. Conf. Soil Mechanics, v. 1, p. 192-198.

Searight, T. K., 1959, Post-Cheltenham Pennsylvanian stratigraphy
of the Columbia-Hannibal region Missouri: unpublished Doctoral
dissertation, Univ. of Illinois, Urbana, Illinois, 256 p.

Searight, W. V., 1955, Guidebook, Field Trip, Second Annual
Meeting, Assoc. of Missouri Geologists: Missouri Geol. Survey,
Rept. of Inv. 20, 44 p.

———, 1955, Stratigraphic pattern in pre-Marmaton Desmoinesian
cycles (abstract): Geol. Soc. America Bull., v. 66, no. 12, pt. 2, p.
1614.

———, 1958, Pennsylvanian (Desmoinesian) of Missouri: Guidebook,
Field Trip no. 5, St. Louis Meeting, Geol. Soc. America, 46 p.

———, 1961, Pennsylvanian System, *in* Howe, W. B., and J. W.
Koenig, The stratigraphic succession in Missouri: Missouri Geol.
Survey, v. XL, 2nd ser., p. 78-95.

——— & N. W. Jeffries, 1958, Alvis and Lexington coals of Missouri
and associated beds (abstract): Geol. Soc. Am. Bull., v. 69, no.
12, pt. 2, p. 1641.

———, W. B. Howe, R. C. Moore, J. M. Jewett, G. E. Condra, M.
C. Oakes, & C. C. Branson, 1953, Classification of Desmoinesian
(Pennsylvanian) of northern Mid-continent: Am. Assoc. Petroleum
Geologists Bull., v. 37, no. 12, p. 2747-2749.

——— & T. K. Searight, 1961, Pennsylvanian geology of the Lincoln
fold: Guidebook, 26th Annual Field Conference, Kansas Geol.
Soc., p. 156-163.

Shaw, A. B., 1964, Time in stratigraphy: McGraw-Hill Book Co.,
363 p.

Sholten, Robert, 1959, Synchronous highs: Preferential habitat of
oil: Am. Assoc. Petroleum Geologists, v. 43, no. 8, p. 1793-1834.

SIEVER, RAYMOND, 1957, Pennsylvanian sandstones of the Eastern Interior basin: Jour. Sedimentary Petrology, v. 27, no. 3, p. 227-250.

SPIEKER, E. M., 1946, Late Mesozoic and Early Cenozoic history of central Utah: U.S. Geol. Survey, Prof. Paper 205-D.

STRACHAN, C. G., 1957, The interpretation of structure in exploration for oil and gas: Tulsa Geol. Soc. Digest, v. 25, p. 132-134.

TEAS, L. P., 1923, Differential compaction the cause of certain Claiborne dips: Am. Assoc. Petroleum Geologists Bull., v. 7, p. 370-378.

TENNISSEN, A. C., 1967, Clay mineralogy and ceramic properties of lower Cabaniss underclays in western Missouri: Missouri Geol. Survey, Rept. Inv. 36, 55 p.

TERZAGHI, R. D., 1940, Compaction of lime mud as a cause of secondary structure: Jour. Sedimentary Petrology, v. 10, p. 78-90.

TRASK, P. D., 1931, Compaction of sediments: Am. Assoc. Petroleum Geologists Bull., v. 15, p. 271-276.

VAN TUYL, F. M., & B. H. PARKER, 1941, The time of origin and accumulation of petroleum: Quarterly Colorado School of Mines, v. 36, Chap. 11 (Recurrent folding and accumulation), p. 83-89.

WANLESS, H. R., 1931, Pennsylvanian cycles in western Illinois: Illinois Geol. Survey, Bull. 60, p. 179-193.

———, 1964, Local and regional factors in Pennsylvanian cyclic sedimentation, *in* Merrian, D. F., Symposium on cyclic sedimentation: Kansas Geol. Survey Bull. 169, v. 2, p. 593-606.

——— & J. M. WELLER, 1932, Correlation and extent of Pennsylvanian cyclothems: Geol. Soc. America Bull., v. 43, no. 12, p. 1003-1016.

——— & F. P. SHEPARD, 1936, Sea level and climatic changes related to late Paleozoic cycles: Geol. Soc. America Bull., v. 47, no. 8, p. 1177-1206.

———, 1947, Regional variations in Pennsylvanian lithology: Jour. Geol., v. 55, no. 3, pt. 2, p. 237-253.

——— and others, 1963, Mapping sedimentary environments of Pennsylvanian cycles: Geol. Soc. America Bull., v. 74, no. 4, p. 437-486.

——— and others, 1965, Late Paleozoic deltas in the central and eastern United States (Abstract): Amer. Assoc. Petroleum Geologists Bull., v. 49, no. 3, pt. 1, p. 362.

WELLER, J. M., 1930, Cyclical sedimentation of the Pennsylvanian Period and its significance: Jour. Geol., v. 38, no. 2, p. 97-135.

———, 1931, The conception of cyclical sedimentation during the Pennsylvanian Period: Illinois Geol. Survey, Bull. 60, p. 163-177.

———, 1956, Argument for diastrophic control of late Paleozoic cyclothems: Am. Assoc. Petroleum Geologists Bull., v. 40, p. 17-50.

———, 1957, Paleoecology of the Pennsylvanian Period in Illinois and adjacent states, *in* Ladd, H. S., ed., Treatise on marine ecology and paleoecology, pt. 2, paleoecology: Geol. Soc. America, Mem. 67, p. 325-364.

————, 1958, Cyclothems of larger sedimentary cycles of the Pennsylvanian: Jour. Geology, v. 66, p. 195-207.

————, 1960, Stratigraphic principles and practice: Harper and Brothers, 725 p.

WHEELER, H. E., & H. H. MURRAY, 1957, Base-level control patterns in cyclothemic sedimentation: Am. Assoc. Petroleum Geologists Bull., v. 41, no. 9, p. 1985-2011.

WILLIAMS, E. G., & J. C. FERM, 1964, Sedimentary facies in the lower Allegheny Rocks of western Pennsylvania: Jour. Sedimentary Petrology, v. 34, p. 610-614.

ZANGERL, RAINER, & E. S. RICHARDSON, 1963, The paleoecological history of two Pennsylvanian black shales: Chicago Natural History Museum, Fieldiana, Geology Mem., v. 4, 352 p.

≋ About the Author

RICHARD J. GENTILE has served the Missouri Geological Survey and Water Resources Division as Geologist (1958-1966) and Chief Geologist and Head of the Coal Geology Section (1966).

He prepared for his profession on the campuses of the University of Missouri at Columbia (B.A., 1956; M.A., 1958) and at Rolla (Ph.D., 1965). He has done extensive field mapping of the rock systems in Missouri, and his work is included in reports of the Missouri Geological Survey and other institutions. Dr. Gentile spent the summer of 1963 at the Gulf Coast Research Laboratory at Biloxi, Mississippi, where he studied present-day environments of sediment deposition. He also has pursued advanced studies in coastal processes and paleogeography at Florida State University, Tallahassee. Observations of present-day geological processes in the Gulf Coastal Area have been applied to the interpretation of the rock sequence in western Missouri.

Dr. Gentile is at present Assistant Professor in the Department of Geology and Geography in the University of Missouri at Kansas City.